Plumbing:

A guide for the Illinois Apprentice Plumber

Second Edition

By
Reese Menard

Illinois Plumbing Seminars LLC

Copyright © 2016 Reese Menard
All rights reserved.
ISBN 10:1539569667
ISBN-13:9781539569664

DEDICATION

I would not have been able to write one sentence of this book without the support and guidance from the following people: First, my lovely wife Chrissy. She has been by my side since the very beginning of my career. Her encouragement and support has been my foundation. My two children, Brandon and Alexis; who have inspired and encouraged me during the creation of this book. I also need to thank and acknowledge the people who have help me to develop my career on a technical and professional level. Thank you to all of you I have met along the way.

FORWARD

The book in your hands is the first of its kind within the Illinois plumbing world. What I mean by that is until now there has never been a study guide for the Illinois apprentice plumber. Currently and for more than 20 years that I know of, the Illinois plumbers' exam has been a 70 percent fail rate.

Why is there such a high rate of failure? Several reasons, first I will state the obvious preparation, or lack thereof. Secondly let us be honest this is a difficult exam. Even after all the years ago that I had taken the exam, and passed it the very first time, something that I am proud of; I can still recall the level of anxiety over the test day. Standing in a line up with 100 other applicants' shoulder-to-shoulder nervously awaiting instructions.

Yes, I am a plumber. An honest to goodness full-fledged Illinois State licensed plumber. I started out in this industry purely by accident. In 1986 at the ripe old age of 19, I did what most would consider being unimaginable. I got married to my high school sweetheart. However, with that commitment, I had a problem.

How was I going to support a wife? At the same time, I was working part time in a warehouse and attending class part time at a community college studying to become the next Frank Lloyd Wright. I had dreams of following in the footsteps of one of the awe-inspiring architects of the twentieth century. As fate would have it one Friday night, I received a phone call from a friend asking if I would be interested in helping him out on Saturday on a plumbing job in the City.

Desperate for money, I jumped at the chance! I immediately enjoyed the work; and the pay was not so bad either. One thing led to another and I dropped out of college. I began working with a large residential plumbing company in the suburbs. The first year of my career had been spent primarily on the business end of a shovel. I started out installing subsoil drain tile, sump crocks, and the occasional bathroom underground.

I had not begun to accumulate time towards my apprenticeship until 1989, at the time I was working for a plumbing service shop in the suburbs of Chicago. I had gone from digging ditches on a new construction residential site to a service shop just that quick.

However, what I really wanted to do was work on the interior plumbing system. It was difficult to get a position as an apprentice plumber back then. Therefore, I did what anyone else would do; I moved on to a small plumbing company that would help me get to the next level. For the next two years, I ended up working on the drain cleaning equipment, commonly referred to as a rodding machine.

From 1987, until 1989, I did a lot of drain cleaning and for the most part, I worked as a plumber's helper. I do not regret it at all; I did get a lot of experience rodding sewer lines, kitchen sink, and bathtubs. Then finally, in 1989 I received the break I was looking for. I met Ben the owner of a plumbing, sewer, and water contracting company. Ben in turn introduced me to Danny, who agreed to sponsor me under his license. Now Danny had a Chicago license however he was able to sponsor me through the IDPH, (Illinois Department of Public Health.)
I had filled out the appropriate forms, sent in two pictures and a check for $50.00 dollars (the going rate at the time.)

Three weeks later, I received my apprentice license in the mail. I had also received an 8.5 x 11" green codebook, which I still have to this day. It is still in "like new" condition only because I never really read it. At the time, that codebook appeared to have been written in a foreign language. I could not understand any of it; it was as if written by someone at NASA.
Over the next few years, I had been involved with all types of plumbing. I had worked manly in residential service, which meant I either repaired or replaced a lot of plumbing. I was also involved with new commercial construction, often times "getting it wrong." I will say that in those early years I had really developed my "plumbing chops."
Eventually another few years had gone by when I received a letter from the state letting me know that 60 months had gone by and I could not renew my apprentice card.

It was time to take the state exam!
The hours in between my apprenticeship, were very long, but the work had always motivated me. I wanted to learn everything I could about plumbing.

Working for Ben had allowed just that. For which I am thankful. Most plumbers today have never had a chance to repair or install water main, sewer systems, fire hydrants, or know how to run a pipe-threading machine. I am very happy to have had the opportunities to do so.
I often think of the apprentices who are currently learning this trade, and I think who is going to teach the next generation of people that will ultimately carry on this trade. Who will be the apprentices that will know how to repair a faucet rather than replace it, who will know how to repair a water main when it ruptures?

More than once, I have encountered the "specialty plumber" who can only do one or two aspects of this trade.

My personal thought on the subject is, you can either do all of what is required to do in this field or at the very least learn about all aspects in this trade. I worked as a supervisor for a very large new residential plumbing contractor, and I would have "help" sent out to my job site. It had gotten to be routine to ask whenever a licensed journeyman would show up, "can you install PVC waste and vent?" Repeatedly the answer I would hear is "I'm a water piper" or "I trim."

This over time became the norm. It was then I decided to help apprentices learn all there is to in this trade, or at least the things I have learned.
There are few with the opportunity to have learned how to read a blueprint, layout a building, or figure a take-off for materials. There is so much to know in this trade that a four-year apprenticeship is enough time to prepare for the exam.

The rest of your career you will be spending time to develop the knowledge for advancement, and growth. Anyone who is considering plumbing as a trade my recommendation is to not only purchase an Illinois State Plumbing Codebook but to read it, and study it as well You may not have an understanding of most of it however, between working in the field and reading the codebook repeatedly; things will begin to make sense. This is a highly skilled trade and will require patients to learn and understand it.

Another avenue to gain knowledge is by attending continuing education classes. These are state required classes that all licensed plumbers must take in order to renew their license. They are however NOT required for apprentices to renew. They are however a great way to brush up on new products or codes.

In addition there are several free trade magazines out there and available. The internet is full of good resources; I would advise however, against some video's that are available. I have watched a few and I can say they are not always correct. There are several trade groups you can join; I have provided a websites at the back of this book.

It is for the reasons listed above that has motivated me to write this book. Within these pages, I will share all of the required information you as an Illinois licensed apprentice will need to know to pass the state exam. Preparation is the key to your success in moving your career into the next phase. Back when I was preparing to take the exam, I had very little classroom preparation. Everything I had learned was hands on. Once the day arrived, I felt less than ready.
With little coaching, I set off to take the exam. How is it that everyone forgot to tell me what a debacle the drawing is? I drove three and a half hours back home after the exam just thinking about the exam and having an overwhelming feeling I was going to have to take it again.

Once back to the normal routine of working, after few weeks had gone by and I had forgotten all about the exam. I received a phone call one day while at work from my wife. She was crying and my heart was in my throat. I could not understand what she was saying until she gained her composure and managed to get the words out that I had passed the exam! That was one of the proudest days of my professional life.

It only made sense the next logical thing to do was to open my own plumbing shop. For a word of caution, you maybe the best technical plumber out there but a little business sense really goes along way!
I was not the latter. I could repair, install, or even design just about any plumbing project that came along. However, I had very little to no business management skills to brag about.
The company I had formed was out of business four years after it started. It was then that I realized I could work for a company but I could not do it on my own in my business. The reason was simple; I focused on the work and not the business side of the operation. A few more years went by and I decided to become a plumbing instructor, while at the same time I had become a plumbing manager for a national company.

My business skills improved and my teaching career began. I also became a continuing education sponsor. I really enjoy teaching apprentices. It is amazing to see the look on an apprentices' face when they fully understand a certain topic or system.

Teaching is a fantastic experience. *Docendo discimus*, which is Latin for "by teaching we learn," ring true; teaching allows us to be much more receptive to learning and to desire moving through the process of novice to expert and into mastering a subject.
Teaching is also about sharing knowledge. When knowledge is not passed on, it becomes lost. In addition, this trade will surprise you. Just when you think you have figured it out and have it mastered, you just might learn something new when you least suspect it.

I also have experience working as a plumbing inspector. In 2013, I decided to apply for and successfully pass the state exam to become a certified plumbing inspector. In addition, I am currently employed as such for multiple municipalities.

As of 2016, there are approximately 8,700 licensed plumbers in Illinois with a state license and roughly 4,000 with a City of Chicago license.
There are about 13,000 plumbers for a state with a population of roughly 12.9 million people. Chicago alone has a population of about
2.6 million. Is that a large number people's health to safeguard?

The bottom line is our trade has been dropping in size for a number of years. There is a decreased desire to work in the trades, let alone a dirty trade. The demand is growing for tradesmen according to labor board statistics. As of 2014, the plumbing trade is expected to grow by 12 percent over the next ten years- faster than the average for all other occupations.

Plumbing has been a good trade to my family and me. There have been difficulties along the way, but this trade has always provided me with the means to raise a family, buy a home, and live my version of the American dream.

My hope is that within this book I have provided some encouragement to you as a licensed apprentice, or if you are considering plumbing as a trade. Plumbing can be a highly lucrative and fulfilling career path. It is with that I hope you enjoy, learn, and retain the information within the pages of this book. In addition, my goal in creating this book is for you to pass the state exam! Best of luck to you in your career as an Illinois state licensed plumber.

CONTENTS

PART ONE: THE BASICS

1 HOW DID WE GET HERE

The Roman Empire
The Middle Ages
Plumbing begins to evolve
Modern plumbing
Death in the water
The process
Death in the air
Trouble in Milwaukee
Where it all ends up
Worldwide water

2 NAVIGATION

Old VS. New
Illinois plumbing license law
Permits and inspections

3 PLUMBING DEFINITIONS AND TERMS

Common definitions used in plumbing

4 PLUMBING MATERIALS

Drain waste and vent
Plastic pipe
Potable water pipe materials

5 JOINTS AND CONNECTIONS

Types of joints used

6 TRAPS AND CLEANOUTS

What is a trap?
Trap types
Cleanouts

7 INTERCEPTORS AND BACKWATER VALVES

Types of interceptors
How an interceptor works
Trap sizing
Backwater valves and flood control

8 PLUMBING FIXTURES

The basics
The men responsible
Faucets
Water closet
Bathtubs and whirlpools
Urinals
Mixing valves
Dishwashing equipment
Emergency equipment

9 HANGERS AND SUPPORTS

Types of support systems
Materials and support intervals

10 WATER SUPPLY AND DISTRIBUTION

Water quality
Protecting the water supply
Backflow/what could go wrong
Flushing and disinfection
Water design and sizing

11 HOT WATER SUPPLY

Hot water introduction
Water heaters piping methods
Water heater maintenance

12 DRAIN WASTE AND VENT SYSTEMS

Residential drainage
Drainage fixture units
Hydraulic gradient / vent systems

13 ESTIMATEING PLUMBING WORK

Large and medium projects
Fixture bidding

14 CUSTOMER SERVICE

Plumbing hats
Is the customer always right

PART TWO: EXAMINATION GUIDE

15 EXAM QUALIFICATIONS

Requirements for admission to the exam
Examination and license fees
Definitions
Administration of the exam
Examination results
Licensing of apprentice plumbers
Licensing of plumbers
License records
License violations

16 EXAMINATION STRATEGIES

Test taking strategies
Strategies for multiple choice
State exam required materials
Written exam questions
Plumbing drawings
Plumbing math

SOURCES

USEFULL WEBSITES

FINAL WORD

1 HOW DID WE GET HERE

The Roman Empire

Plumbing has been around for much of human kind. However, safe sanitation practice was not always at the forefront. In fact, it has been usually at the back-end (pun intended!).

Archaeologist have discovered cisterns dating back nearly 6,000 years. A cistern is a fabricated vessel or a storage tank- essentially a well to store and provide water. They can be as deep as ten feet into the earth and are constructed with stone, brick, or adobe.

The Romans are without question the world's first well-known plumbers. They are often referred to as artisans with lead. In fact, the Latin word for lead is *plumbum*. The very word "plumber" dates back to the Roman Empire.

Romans used lead in conduits and drainpipes. Lead was also used in many other things such as bathtubs and early lead water pipes. Today in the city of Chicago, lead pipe water service is still very much in use. In medieval times, anyone who worked with lead had been referred to as a plumber.

Sanitation in ancient Rome was a complex system resembling modern plumbing. The basic law of physics as it applies to plumbing is gravity. Gravity is used to carry wastewater out. Gravity also helps with airflow within a vent system. Gravity in some cases is also responsible for the delivery of fresh potable water into a building.

During the dark ages, the technical knowledge of the Roman system was lost and has been investigated by modern-era historians and archeologists. A system of eleven Roman aqueducts provide the inhabitants of Rome with water of varying quality. The best quality water is reserved for potable supply. The poorer quality of water was used in public latrines and baths. This would be the equivalent to a public washroom of today. Latrine systems have been found in many places such as a Rome and Pompeii.

It is believed that the Romans used sea sponges on a stick dipped in vinegar after defecation, but the practice is a one-and-done use. The Romans also had a complex system of sewers, usually constructed of hand-chiseled stone. Wastewater flushed from the latrines flowed through a central channel and then into the main sewerage system. It is estimated the first sewer systems were built between 800 and 700 BC. Drainage systems evolved slowly, and began primarily as a means to drain marshes and storm runoff. The sewerage system did not really take off until the arrival of *Cloaca Maxima*, an open channel that was later covered, and one of the best- known sanitation artifacts of the ancient world. Most sources believe it was built during the reign of the three *Etruscan kings* in the sixth century B.C.

From early times, the Romans, in imitation of the Etruscans, built underground channels to drain rainwater that might otherwise wash away precious top soil. Over time, the Romans expanded the network of sewers that ran through the city and linked most of them together. Including some drains into the Cloaca Maxima, which emptied into the Tiber River. In 33 BC, the Cloaca maxima was enclosed under the emperor Augustus, thereby creating a large tunnel.

Public latrines date back to the 2nd century BC. Whether intentionally or not, they became places to socialize. Long bench like seats with keyhole shaped openings cut in rows offered little to no privacy. Some latrines were free to use while others had fees to use them.

The Aqueducts provided the large volumes of water needed to the city. They had been constructed from the same law of physics that would allow water to drain away from the city- gravity. The aqueducts would carry water from the Anio River. The Anio aqueduct served as our modern day water mains. Aqua Anio Novus, as it was known was a major part of the Roman water system. The water distribution system was carefully designed so that it was separated from the waste system known as Cloaca Maxima.

Other towns and cities in the region copied the system in Rome. Even villas that could afford plumbing had a system to bring fresh water in and carry wastewater away. Roman citizens came to expect high standards of hygiene, and the army was provided with latrines and bathhouses. Water supply and sanitation has been a primary logistic challenge since the dawn of civilization.

Where water resources, infrastructure, and sanitation systems are insufficient for the population, people fall prey to disease, dehydration, and possible death. Major human settlements could only initially develop where surface water was plentiful, such as near rivers, streams.

Over the millennia, technology has dramatically increased the distances across which water can be relocated. However, the availability of clean fresh water remains a limiting factor on the size and density of the population.

By the 4th century AD, Rome would have 11 public baths, 1,300 fountains and cisterns, and 850 private baths. In Pompeii, some homes had as many as 30 taps. With the decline of the Roman Empire, whose ruination became complete by the 6th century A.D, Roman garrisons in Britain had been invaded by hordes of Saxons, Scots and Irish. They could not count on help from the Rome, which was in trouble itself. When the last Roman garrison fled the Isle of Britain, the secrets of sanitary design went with them.

Replacing them were the Barbarians, leveling cities and decimating populations as they made their way across the continent. Civilization regressed and sanitation technology was reverted to its basic forms.

The Roman Empire would not have been able to construct large cities without plumbing. Their society would have been very different without the ability to have imported water from the Anio River. The city would not have been nearly as clean, nor would they have been referred to as the "bath culture." Visitors to Rome at that time were amazed at the cleanliness. Even today's visitors find the aqueducts to be amazing works of architectural edifice. The arch and the arcade, which is a series of arches, is beautiful, some of the best villas were built to look out over the aqueducts. The other hidden half of their plumbing system were the sewers.

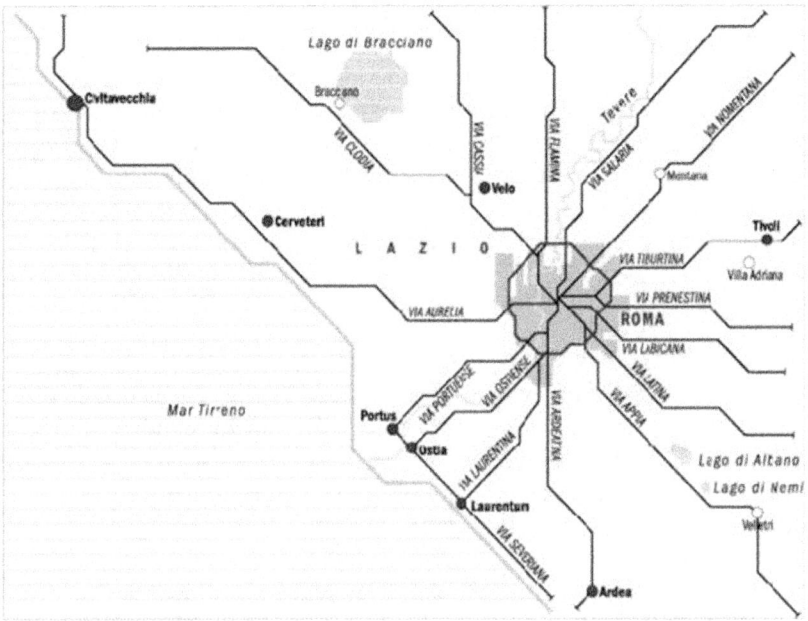

A map of Rome, this map shows the route of 14 aqueducts that would function much as a modern day water main infrastructure. The aqueducts where constructed to bring fresh potable water into the city.

Middle Ages

By the Middle Ages, the "hot houses" or "stews" of the Roman baths carried the stigma of debauchery and wild parties. During the reign of Richard the Lionhearted, the little rooms or "bordellos" of the baths became synonymous with brothels. In 1348, the first wave of Black Plague entered England through the town of Melcombe in Dorset County. One third of the population would be wiped out, as rats and fleas thrived in the filth of garbage and human waste in the streets. It was only a fire in 1366 that had put an end to the Black Plague thereby burning most of England to the ground. Europe, with a rise in population, had its share of sanitation troubles partly because the ill effects of proper sanitation was unknown at that time. Most people had thought bathing was a waste of time. Body and head lice were common among citizens. With the fall of the Roman Empire, sanitation practices had reverted to poor at best. It was common to empty chamber pots directly into the streets.

Plumbing Begins to Evolve

It was not until the 16th century when Sir John Harington, an English author had invented the first flushing toilet. He became a prominent member of Queen Elizabeth's court. Harington's invention was, at the time called the Ajax, or "Jakes" as a slang term for toilet. Harington's design a forerunner to the modern day flush toilet had a flush valve to let water out of the tank, and a wash-down design to empty the bowl. The term "John", used particularly in the United States, is generally accepted as a direct reference to its inventor.
It would not be possible to write a section of any plumbing history without talking about Thomas Crapper. Thomas Crapper all jokes aside was in fact a plumber who lived in London from 1836 – 1910. The owner and founder of Thomas Crapper & CO. One large misconception is, he was the inventor of the flush toilet, we now know, as previously talked about in the last paragraph is not true. Crapper however did own and operate the first bath, toilet and sink showroom.
Crapper also greatly improved upon much of the design's for plumbing fixtures. Manhole covers that bear Crapper's company name on them in Westminster Abby are now one of London's minor tourist attractions. Crapper served his apprenticeship under his brother George, who was a master plumber.

In 1861, Crapper set himself up as a sanitary engineer, with his own brass shop and foundry.
Crapper heavily promoted sanitary plumbing and pioneered the concept of the bathroom fittings and showroom. Prince Edward purchased crapper's plumbing fixtures.

An order that included thirty of Thomas Crapper's lavatories, with cedar wood seats and enclosures, thus giving Crapper his first Royal Warrant. The firm received further warrants from Edward as king and George V both as Prince of Wales and as king.

In 1904 Crapper retired, passing the firm onto his nephew George and his business partner Robert Marr Wharam, Thomas Crapper died in 1910.

In 1966 the company was sold by the owner Robert G. Wharam, son of Robert Marr Wharam, to their rivals John Bolding & Sons, Bolding went into liquidation in 1969. During Thomas Crapper's time as a plumber, he held nine patens, three of them for improvements to the water closet. The company is still in business to this day and is still manufacturing bathroom-plumbing fixtures.

Modern Plumbing

Plumbing has come a long way from the days of aqueducts, to chamber pots, and the modern flush toilet, as we know it. The executive mansion, also known as the White House, had not been upgraded and fitted with indoor plumbing until the frustration Mrs. Roosevelt had put up with. Our very own White House did not have plumbing installed until 1902, the first of many remodeling projects. The price tag for the remodel a whopping $500,000.

A big influence in the change of sanitation and plumbing was brought on by the invention of the microscope in 1590. This new invention had allowed doctors of that time to visually sample water. They had been able to see waterborne disease for the first time. We will look at three separate cases where this is evident. The largest documented case in US history was in Milwaukee, Wisconsin back in 1993. We will exam that case further at the end of this chapter.

Plumbing, like other industries has evolved over the years that is with the help of people like Thomas Crapper, John Harington, and John Kohler. These are just three who have with hard work and dedication help change and better this industry.

Without properly installed plumbing, as history shows us, filth would run through the streets, disease would run rampant, decimating populations across the world.

In 1873, John Kohler looked at a cast iron horse trough and realized it could also serve as a way for people to be able to bathe. Thus, the cast iron bathtub was created. Kohler, of course is one of the best known, and largest manufactures of plumbing fixtures.

American Standard is another manufacture of plumbing fixtures. They have been in business for the last 140 years as well. John B. Pierce, the founder of American Standard, started much the same way as Mr. Kohler.
American Standard used to be known as Standard, up until about the 1920's. The rumor is that at a trade show in or around 1926; the slogan for "plumber protects the health of the nation" had first appeared and was where the original posters had been used at that trade show. For me it is far more than just a slogan, it is the truth, and history has taught us the lesson. As you are just beginning you career in plumbing, always remember that phrase. Think about this also; plumbing is the highest regulated trade among all construction trades, and for good reasons.

Modern society regards an uncouth person as one who does not wash and deodorize every day. Yet this is a custom of relatively recent vintage. As recently as the 1950s, many working-class Americans still lived in cold-water flats without a bathtub. For persons who grew up in humble circumstances, bathing once a week was the norm.
Traditionally, bathing would take place on Saturday night in order to be clean for church services the next day. Daily bathing was simply too inconvenient, more a luxury than a necessity.

As the first three decades of this century progressed, tubs came to be installed in virtually all new dwellings and retrofitted in many others. The trend was slowed by the Depression and World War II but resumed at a hastened pace during the era of postwar prosperity. Now, of course, the basic question asked by prospective homeowners and renters is not whether the dwelling has bathing facilities, but "how many?"

Cultural acceptance of daily bathing was sparked in great measure by an annual "Bath a Day" campaign mounted by the plumbing industry to promote sales of bathtubs and related products. The campaign was originated in 1914 by the old Domestic Engineering magazine. A few years later, it was adopted and expanded by the Trade Extension Bureau, forerunner of today's Plumbing-Heating-Cooling Information Bureau, which came into existence in 1919. The soap industry soon joined in with an allied effort of extensive consumer promotion.

"Bath a Day" was sustained throughout the 1920s with a variety of consumer promotions, quite successfully it seems. Consider this editorial from the New York World, Dec. 13, 1929. The time is soon coming, according to the computation of those attending the soap convention in Chicago, when every night will be Saturday night in the American home, and a bath every day will be taken as a matter of course by the completely American nation. Already, it appears, we consume 3 billion pounds of soap per annum, which would indicate a bath two or three times a week for everybody and the total tends to rise.

Much of the credit, it seems to us, will go to the manufacturers of bathroom fixtures. The strides they have made in the last 30 years are hardly to be described in prose. As late as the Spanish-American War, most Americans, if they had bathrooms at all, were provided with tubs made out of some shiny metal, probably tin, and with other fixtures of the same general pattern. Obtaining hot water was extremely difficult, involving as it did a deal of yelling to Sarah, down in the kitchen, to start the heater going, or else requiring the operation of a smelly heater in the bathroom itself. Now, however, all that is changed. In the big cities, even the cheapest apartment has a bathroom that would put the best bathrooms 30 years ago to shame. The fixtures are porcelain, or of metal enameled in excellent imitation of porcelain; the hot water runs hot water; the floor is tiled; and so are the walls.
Death in the Water

Imagine what life would be like without a modern plumbing system. How would you cope without running water or flushing toilets? Water is the most essential nutrient to life on earth; it forms the basis of a healthy diet and lifestyle.

The world today with its large, high-density population would not sustain life, as we know without a modern plumbing system that provides clean drinking water and sewage systems that remove contaminated used water.

The past is filled with historical events that show successes and failures of plumbing. One of the most recent examples is Flint, Michigan, whose citizens were hit among the hardest during the housing market crash of 2008. Which has the citizens of Flint saying to themselves

How did we get here? (Case number one), the timeline of events are as follows.

April 25, 2014
The city switches its water supply from Detroit's system to the Flint River. The switch was made as a cost-saving measure for the struggling community. Almost immediately after, residents begin to complain about the water's color, taste, and odor. Reports of rashes and concerns about bacteria develop.

August & September 2014
City officials issue boil-water advisories after coliform bacteria are confirmed in tap water.

October 2014
The Michigan Department of Environmental Quality blames cold weather, aging pipes, and a population decline. They have been quoted as stating, "The city has taken operational steps to limit the potential for a boil-water advisory to re-occur"

October 2014
General Motors plant in Flint stops using municipal water, saying it corrodes car parts.

January 2015
Detroit's water system offers to reconnect to Flint, waiving a $4 million dollar connection fee. Three weeks later, Flint's state-appointed emergency manager, Jerry Ambrose, declines the offer.

February 2015
In a memo for the governor, officials play down the problems and say that the water is not imminent "threat to the public health."
"It's clear the nature of the threat was communicated poorly. It's also clear that folks in Flint are concerned about other aspects of their water; taste, smell and color being among the top complaints."

February 18, 2015
114 parts per billion of lead are detected in drinking water at the home of Anne Walters. Mrs. Walters notifies the Environmental Protection Agency. Even small amounts of lead can cause lasting health and development problems in children. The E.P.A. does not require action until levels reach 15 parts per billion, but public health scientist say there is no safe level for lead in water.

February 26, 2015
A water expert from the E.P.A. and the Department of Environmental Quality discuss high levels of lead found in a resident's water sample.

February 27, 2015
Miguel Del Toral, an E.P.A. expert, says that the state was testing the water in a way that it could profoundly understate the lead levels. "Given the very high lead levels found at one home and the pre- flushing happing in Flint, I'm worried that the whole town may have much higher lead levels than the compliance results indicated."

March 3, 2015
Second testing detects 397 parts per billion of lead in drinking water at Mrs. Walters home.

March 12, 2015
Veolia, a consultant group hired by Flint, reports that the city's water meets state and federal standards; it does not report specifically on lead levels.

May 6, 2015
Test reveal high lead levels in two more homes.

July 2, 2015
An E.P.A. administrator tells Flint's mayor that "it would be premature to draw any conclusions" based on a leaked internal E.P.A. memo regarding lead.

July 22, 2015
Dennis Muchmore, Governor Rick Snyder's chief of staff, expresses concern about the lead issue in an email, and asks about Flint test results, blood testing, and the state's response.

August 17, 2015
Based on results showing lead levels at 11 part per billion from January to June 2015, the Department of Environmental Quality tells Flint to optimize corrosion control.

September 2, 2015
Marc Edwards, an expert on municipal water quality and professor at Virginia Tech, reports that corrosiveness of water is causing lead to leach into the supply. Soon after, the Department of Environmental Quality disputes those conclusions.

September 24-25, 2015
A group of doctors led by Dr. Mona Hanna-Attisha of Hurley Medical Center in Flint urges the city to stop using the Flint River for water after finding high levels of lead in the blood of children. State regulators insist the water is safe.

"D.E.Q. and D.C.H. feel that some in Flint are taking the very sensitive issue of children's exposure to lead and trying to turn it into a political football claiming the impacts on the populations and particularly trying to shift responsibility to the state."

September 28, 2015
The governor is briefed on lead problems in a call with the state environmental department and federal officials.

October 1, 2015
Flint city officials urge residents to stop drinking water after government epidemiologists validate Dr. Hanna-Attish's finding of high lead levels. Mr. Snyder orders the distribution of water filters, the testing of water in schools and the expansion of water, and blood testing.

October 16, 2015
Flint reconnects to Detroit's water supply. Residents are advised not to use unfiltered tap water for drinking, cooking, or bathing.

October 19, 2015
The Department of Environmental Quality director, Dan Wyant, reports that his staff used inappropriate federal protocol for corrosion control.

October 21, 2015
Mr. Snyder announces that an independent advisory task force will review water use and testing in Flint.

December 9, 2015
Flint adds additional corrosion controls.

December 14, 2015
Flint declares an emergency.

December 29, 2015
The task force says the Department of Environmental Quality must be held accountable. Mr. Wyant, the director of the state environmental agency, resigns. "We believe the primary responsibility for what happened in Flint rests with the Michigan Department of Environmental Quality."

January 5, 2016
Mr. Snyder declares a state of emergency for Genesee County, which includes Flint. The Department of Justice opens an investigation into the issue.

January 12, 2016
Governor Snyder calls out to the National Guard to distribute bottled water and filters in Flint.

January 13, 2016
The crisis expands to include Legionnaires' disease. Officials reveal a spike in cases, including 10 deaths, after the city started using Flint River water.

January 16, 2016
President Obama declares a state of emergency in the city and surrounding county, allowing the federal Emergency Management Agency to provide up to $5 million in aid.

January 20, 2016
Governor Snyder releases more than 270 pages of emails about the Flint water crisis that show debate over who is to blame and offer insight into the state's response – the first batch of several that are rolled out in the coming months. President Obama says he "would be beside myself" if he were a parent in the city of Flint. The Michigan House approves $28 million requested by the governor to assist the city.

February 2016
Several lawsuits are filed over the lead-tainted water crisis, including some that name Governor Snyder and public employees.

February 2016
Plumbers from United Association Local 370 in Flint have been volunteering since October 2015 to install filters and faucets in order to help get lead out of people's tap water. More than 300 plumbers driving in from Lansing, Detroit, Saginaw, and other cities across Michigan. Armed with donated faucets and filters; provide by such companies as Moen, Brasscraft, Speakman, Delta, and American Standard. The plumbers descend on Flint.

March 2016
Snyder, the state appointed emergency manager who oversaw Flint when the water source was switched to the river and other state officials testify in front of Congress.

March 23, 2016
A governor appointed panel concludes that the state of Michigan is "fundamentally accountable" for the crisis because of decisions made by environmental regulators.

April 20, 2016

Two state officials and a local official are charged with evidence tampering and other crimes in the Michigan general's investigation, the first to be levied in a probe expected to expand.

June 2016/summary of events

The Flint water crisis began with an April 2014 decision by a state-appointed emergency manager to switch the city's water source from Lake Huron water to a local corrosive river. The state's environmental agency oversaw the switch and failed to require the use of corrosion control agents, which allowed lead to leach off water pipes and flow into households across the city. The glaring oversight was not revealed until last summer. State officials have yet to declare unfiltered tap water in the city safe for consumption, and an investigation into the possible connection between the interim water source and a massive outbreak of Legionnaires' disease – including 12 deaths – remains ongoing.

The projected $216 million to fix infrastructure is just one of the litany of costs inherited by federal, state, and local agencies in the wake of public health emergencies. Some lawmakers have projected total costs including dissemination of bottled water, lead testing, and health treatment at $1 billion.

The city and state also face $1 billion in potential legal liability. A recent state analysis found that monthly water bills for Flint residents – already among the highest in the US – are projected to double over the next several years. However, the report reiterated a problem that has yet to be fully resolved: it is unclear exactly how many lead services line are in the city, which could raise the total cost even further. There are no records for about 9,000 lead and galvanized services, the report estimated, and "additional costs may be incurred if all galvanized services are to be replaced".

The governor's office has kept an optimistic view on the situation, however. Snyder aides who are assisting in efforts to address the crisis believe Weaver's estimate of $55m may be sound.

The projected cost to repair infrastructure after the city of Flint, Michigan's two-year water contamination crisis is several magnitudes higher than what has been allocated to fix it, a new state report has found.

The report lays out a bruising litany of infrastructure fixes to the city's water system over the next several decades at an estimated cost of at least $216 million. The report suggests $80 million is needed to remove about 10,000 lead pipes across the city – more than three times what Michigan Governor Rick Snyder has proposed for a forthcoming state budget.

The report from Flint-based engineering firm Rowe Professional Services calls for the widely supported removal of lead pipes in the city to be completed in eight years. The city's mayor, Karen Weaver, has estimated $55 million is needed to remove the pipes, and as many as 500 could be removed during an initial phase launched with $2 million from the state.

The floodgates have been opened so to speak-
With all eyes on Flint Michigan, the lawsuits have begun to flow in Chicago as well. In 2013, a 24-page complaint found high levels of lead in the drinking water in homes. This after the claims of recent construction in the area had disrupted lead service lines.

There is no evidence that Chicago faces problems of the same magnitude as Flint. However, unlike many other U.S. cities that shifted decades ago to pipes made of copper or other metals, Chicago required the use of lead service lines until they were banned nationwide in 1986. Older cities, including Chicago, add corrosion-fighting chemicals to the water supply that form a protective coating inside pipes.
Officials in Flint stopped the treatment in an ill-advised attempt to cut costs. EPA study and other research found the anti-corrosion treatment also might be ineffective when street work, plumbing repairs, or changes in water chemistry disrupt the coating, causing lead to leach from service lines. The coating used to prevent leaching is phosphate; orthophosphate is widely used for this reason. The U.S. Food and Drug Administration and the EPA recognize it as safe.
Roughly 10 million American homes and buildings receive water from service lines that are at least partially lead, according to the Environmental Protection Agency. Service lines are the pipes connecting water mains to people's houses. Lead ones are mostly found in the Midwest and Northeast.

Despite the life-altering consequences of lead poisoning, there is no national plan to get rid of those pipes.

A top reason for continuing to use lead service lines instead of immediately digging them up is that utilities can treat water so it forms a coating on the interior of the pipes — a corrosion barrier that helps prevent lead particles from dislodging and traveling to your faucet.
However, if the water chemistry changes, the corrosion controls can fail.

Even replacing the lines can be trouble, however, as the law only requires replacing the lines on public property — replacing the portion of a lead service line on private property is up to the owner — and it turns out that replacing just the public portion of a lead service line can cause lead levels to spike in a homeowner's water. That is because the work involved in replacing just part of a lead service line can jostle free lead in the remaining part of the pipe.

The Process

Across the city of Chicago, thousands of residences are connected to lead pipes, including water services and water mains. When street construction or plumbing repairs disrupts pipes, lead is one contaminate that can find its way into the water that flows out of household faucets. In a 2013 study, researchers at the U.S. Environmental Protection Agency discovered what some plumbers already know. First, I would like to address lead contamination in pipping.

How would you know if the water is contaminated with lead? You would not. However, to be clear Chicago follows federal government protocols to test water. There is far less danger for a similar event such as in Flint to occur in Chicago.

For starters, the wide spread lead poising started because the high lead levels in the Flint River. Chicago residents and surround communities receive water from Lake Michigan prior to treatment. Unlike Flint's, water not receiving the correct treatment before making its way to residents.

Chicago does testing in about 50 homes, every three years, which is the federal standard. This by the way is a city of about 5-1/2 million people; homes built before 1986 could have lead pipes carrying domestic water supply everyday into a home. The standard federal rule by the way for testing is to sample the first liter of water drawn in the morning.

The E.P.A. study found that the first liter is often lead free, they did discover high levels of the toxic metal can flow through taps for several minutes afterward, depending in part on the length of the service line between the home and the street.

The federal E.P.A. recommends following these steps to help protect against lead in water.

The water distribution in a plumbing system should be flushed three to five minutes anytime water has not been used for several hours. Which all plumbers do typically after any repair in which the water has been shut off. That flushing of the system can be accomplished by doing a load of laundry.

Secondly, before drinking from the tap flush out the piping for an additional 30 to 60 seconds to clear any remaining water sitting in the homes piping.

Third, the E.P.A. suggest the use of water filtering methods; either a whole house filter or water pitchers equipped with a filtration devise. Chicago's water is clean, safe and abundant, the envy of most other states thanks to the Lake Michigan filtering process.

This brings us to the question of treating water and a look at the process. Lake Michigan provides water to 5.5 million Chicagoans and 118 surrounding suburbs. Approximately one billion gallons of water is treated everyday by way of the world's largest water treatment plant- the plant located just north of Navy Pier, known as the James W. Jardine water purification plant. The plant had opened in 1968 and provides virtually all of Chicago's potable water.

The process begins about 2.5 miles off shore at massive in flow points called cribs. Each crib pulls water from 20 feet below the surface down a 168-foot vertical shaft to tubes cut from the bedrock, which flow back to the shore. Once it reaches the plant, fish and other debris are filtered out using a rotating screen and the water is pumped 25 feet up to begin the gravity – powered treatment process.

The water is first chemically treated with chlorine to kill off microbes and activate carbon to remove foreign odor and taste, then fluoride is added, and finally aluminum sulfate about an hour into the process to increase the stickiness of microscopic solids which then adhere to each other, creating floc. Large paddles – floculators – assist with the mixing of aluminum and water.

One last chemical added is polyphosphate, which is used to coat the inside of Chicago's pipes, preventing the lead in old plumbing from leaching into the water supply. The water is then pumped into settling tanks where the floc sinks to the bottom. This sedimentation phase eliminates about 90 percent of the particulate matter from the water. The water is then filtered in one of 96 tanks filled with a layer of coarse gravel under an upper layer of fine sand. Together these layers effectively filter much of the remaining floc and debris from the water.

The entire process from crib to filtration takes eight hours. From here, the water is pumped through 4,000 miles of pipe for consumption.

Replacing lead water services is expensive, and would take years to complete the enormous task. In the meantime, the city can do the next best thing and educate people about the risk and the precautions that can be used. It is also the plumbers' responsibility to do the same.

After all, the plumber is the first line of defense, and ultimately the face of protection against things such as lead poisoning; and will see more people every day.

What the researchers had found in 2013 in Chicago is that alarming levels of lead can flow from the very faucets we drink from. What most people do not know is that anytime a plumbing system, water supply, or distribution is shut off for a repair there is a certain fluctuation that occurs.

This is obvious during a shut down. However, this unstable pressure change will disrupt a thin layer of micro-slime (biofilm) from inside the walls of pipe.

Death in the Air

The same can hold true for Legionella, (Case number two) which is ever present in all plumbing connected to a hot water source. Now an interesting note, we know Legionella is in all plumbing systems and when we drink the water tainted with Legionella, we do not become sick. Its only when the disease is in mist form that it can it become deadly such as through water cooling towers, fountains, humidifiers. Just about any type of warm water in mist form will suffice. As for plumbing in the news, Legionella or Legionnaires disease has also made headlines in the last few years.

Legionella was first discovered in December of 1976. 5 months after the outbreak. Dr. Joseph McDade a CDC laboratory scientist, using the technique of a guinea pig inoculation, was able to isolate the bacterium that caused the disease and identified as Legionella pneumophila.

Legionella was discovered after an outbreak in Philadelphia at an American Legion convention. Those in attendance were mostly veterans of World War II. From July 21 to the 24th, more than 600 Legionaries were staying at the Bellevue-Stratford Hotel.

The disease claimed the lives of 29 people and nearly 200 others sickened by this new outbreak. Shortly after the last day of the meeting, several people became sick with symptoms of chest pains, and high fevers, and died not long after.

They had spent time all over the city, eaten at restaurants, and traveled home in a variety of ways. For the first time, many Americans became aware of the work the Center for Disease Control did. At that time, an obscure organization went to work trying to figure out what was killing the Legionaries. In some ways, the "detectives" working the case raised more questions than answers. A check of the hotels at which the convention goers had stayed revealed no outbreak of the mysterious illness among employees who had been exposed to the Legionnaires.

The investigators found no evidence that any of the victims had been exposed to pigs, which have been implicated as animal reservoir for the swine flu virus. Nor could the "disease detectives" explain another apparent contradiction why some people developed the disease, while others, who ate the same meals, drank the drinks or shared their rooms during the convention did not.
The disease remained elusive. "There's an outside chance we may never find out the cause," CDC Director David Sencer told Time magazine during an interview. "There are times when a disease baffles us all. It may be sporadic, a onetime appearance."

Speculations ran wild some thought that domestic terrorists through chemical or microbiological means caused the epidemic; others blamed the CIA in an experiment that had gone wrong. One Philadelphia toxicologist thought that the epidemic was due to nickel carbonyl poising. Other speculations included toxic fumes from a copy machine, and even an attack by extra-terrestrial forces.

Eventually the disease tapered off and, though nearly 200 had become ill and 29 had died, fear also tapered off. It was clear that whatever it was, it was not spreading – but though public interest waned the CDC kept working. Eventually the source of the disease was traced back to the hotels air conditioning unit. The epidemic caused such a stir that congress had held hearings leading to the demise of the Bellevue – Stratford Hotel, closing its door permanently.

Chicago had a Legionella scare back in September 2013; The JW Marriott Hotel Chicago had removed a fountain located in the lobby and closed parts of the spa, after three Legionella related fatalities occurred. Illinois Department of Public Health had confirmed ten Legionella cases and three deaths because of people breathing in the bacteria latent water droplets. There are a reported 30 cases a year of Legionella in Chicago alone.

Legionella can be prevented. In the original case from Philadelphia, routine maintenance could have helped prevent the outbreak. The filters on most cooling towers require cleaning and or replacement. The bacteria is present in water with temperatures ranging from 68 degrees to 120 degrees.

Once again, just to be clear, Legionella is present in all plumbing systems. It is dormant in water temperatures in the range of 32 degrees and 68 degrees. However, it cannot survive above 140 degrees it quickly dies off. As plumbers, what we should be aware of is; high water temperature is the easiest and most effective way to protect the public from this disease.

I know you veteran plumbers that may be reading this and are thinking; sure but the Illinois State Plumbing Code Book says water temperatures cannot exceed 110 at a public lavatory, and cannot exceed 115 at any bathtub or shower. Well you would be correct! However if the proper safe guards are in place you can achieve both the satisfaction of the plumbing code and helping to kill off the bacteria.

One thing to remember is that there are several safeguarding devices and many manufactures of thermal mixing valves available. In addition, a recirculating pump must be used continuously within the hot water distribution.

One other way to help prevent bacteria in plumbing is as simple as the removal off "dead ends" within the piping system. Our plumbing codebook tells us any section of dead piping exceeding 24" is to be avoided. We have three major things to consider when installing plumbing as to safeguard against bacteria in the system.

1. Scale and sediment supply the environment needed for growth of Legionella disease bacteria (LDB) and other microorganisms.

2. Dead ends and non-recirculated plumbing lines that allow hot water to stagnate also provide areas for growth of the organism.

3. Temperatures maintained below 60 degrees encourage growth of LDB and other microorganisms.

With that, I also would need to include the three components to aid in system controls.

1. Run recirculation pumps continuously and exclude them from energy conservation measures.

2. Eliminate or minimize the use of rubber, plastic, and silicone gaskets in the plumbing system. These materials serve as growth substrates for LDB. Frequent flushing of these lines also reduce growth.

2. Identify and test the integrity of all backflow preventers (to ensure protection of domestic water from cross-contamination with process water) through building code-approved method.

Yet there are alternative methods to consider in order helping eliminate the possibility of Legionella from the plumbing system; however, the high water temperature as previously discussed is the best. UV or ultraviolet radiation systems are one alternative, dosing with chlorine, hydrogen peroxide is another.

When dosing a system there is, a fair amount of monitoring that is required. Too much chemical additive and you are worse off than before, too little and the effect becomes futile. In addition, some other disadvantages of chemical dosing to keep in mind, is a carbon filter is required for dialysis patients.

Point of use filters may be yet another option but just as adding chemicals to potable water there is disadvantages with the use of filters.

The obvious is, it is only suitable at the point of use, must be replaced, and particulates in water may reduce flow. Legionella can be controlled effectively at the design stage of a newly constructed building, as simple as adding a continuous recirculating loop and raising temperatures at the hot water heater to 144 degrees, and then safeguarding against scald.

Trouble in Milwaukee

Waterborne disease outbreaks are relatively rare events in our time (Case number three), but just two decades ago, Milwaukee experienced one of the largest documented drinking water outbreak in US history. Caused by the chlorine – resistant parasite *Cryptosporidium parvum*, the outbreak affected over 400,000 people, 25 percent of Milwaukee's population in 1993. The dollar amount as a result was $96 million in combined healthcare cost and productivity losses.

On Monday morning, April 5 1993, the laboratory's Chief Virologist and Commissioner of Health received calls inquiring about the nature of apparent gastrointestinal (G) illness reported in the city.

The Director of Nursing had hearsay information that some pharmacies were selling out of anti-diarrheal medications. Unknown to the health department at the time, the Milwaukee Water Works had received some complaints regarding the aesthetic quality of the tap water.

After the Commissioner inquired about any information our Public Health Laboratory (MHDL) might have regarding GI illness in the community, we proceeded to call local hospital microbiology laboratories and ultimately emergency rooms and determined there was extreme GI illness throughout the city based the much higher ER patient numbers and increased work load enteric disease tests. Other staff in the department were also seeing similar indicators throughout the city.

From the start, the water was suspect, upon gathering information from the north and south treatment plants via daily water quality reports, from the past 30 days the only changing trend was the increasing turbidity readings from the south plant, yet were within federal limits. The lab had started a survey of city clinical microbiology labs the following day that confirmed a dramatic increase in testing for GI pathogens city wide, yet no agents were identified that would reflect such a widespread illness.

On day three, a local infectious disease physician called the laboratory with a single case of *Cryptosporidium* that fit the profile of ill patients. Based on a few other patients test revealed that *Cryptosporidium* was present in stool samples. By April 7, about 60 hours after the realization of widespread community illness, the Mayor issued a boil water advisory for 880,000 citizens. The Howard Avenue Water Purification Plant was contaminated, and treated water showed turbidity levels well above normal. Turbidity is the cloudiness or haziness of a fluid caused by large numbers of individual particles that are generally visible to the naked eye. Similar to smoke in the air. The measurement of turbidity is a key test of water quality.

In drinking water, the higher the turbidity, level the higher the risk that people may develop gastrointestinal disease. This is especially problematic for immunocompromised people, because contaminates like viruses or bacteria can become attached to the suspended solids.
The official outbreak-related death's was 69, of which 93 percent occurred in persons with AIDS.

Turbidity is commonly treated using either a setting or a filtration process. Depending on the application, chemical reagents will be dosed into the wastewater stream to increase the effectiveness of the settling or filtration process. Potable water treatment plants as in Chicago, often remove turbidity with a combination of sand filtration, settling tanks and clarifiers.

Where it all ends up

As did the Roman Empire, Chicagoans have a unique and complex network of underground piping system in order to remove, and treat the 1.5 gallons of wastewater that is created every day. Once again, the laws of physics take over and allow wastewater to flow downhill into one of the world's largest treatment plant. The Stickney Water Reclamation Plant is 570 acers in Cicero, Illinois. It serves 2.5 million people over a 260 square mile area; including 43 suburban communities.
The plant opened in 1930 and a second (southwest portion) opened in 1939. There are 166 processing tanks; when wastewater enters the plant at one of two locations, the first phase is to physically remove contaminants.

During this first step, "screening" is done to remove debris that could clog the machines downstream of the plant. Non-biological material is washed and then dumped in a sanitary landfill. The water then flows into a primary settling tank where the heavier solids settle to the bottom. The lighter waste floats to the top where revolving skimmers continually collect the solid waste at the bottom and the film from the top.

Phase two of the process begins with microorganisms feeding on the remaining biological waste. This converts the waste into a compound that separates from water. The first set of tanks are aerated to provide sufficient oxygen to allow the microbes to grow, eat, and multiply. A second tank separates the bacteria culture from the treated water. Bacteria clump together and sink to the bottom of the tank. Up to 85 percent of these cultures are recycled to seed the next batch of cleaning tanks. From this point, the water either is discharged into the sanitary and ship canal or if officials want to get the last 5 percent of water clean it moves to the third stage of the process.

Entering the third stage the water is filtered, ammonia and other specific contaminates are removed, and bacteria that could potentially cause fatalities in humans are removed.

The wastewater, now 95 percent clean, can be deposited right back into rivers and streams without any ill effects on the environment; typically, the water is cleaner than the river itself.

The entire process from start to finish is less than 12 hours to complete. All of the waste that has been removed in stage one, also known as sludge or residuals, are first broken down in huge anaerobic digestion tanks.

They are then centrifugally separated from any remaining water, and then dried in large outdoor beds. Then finally hauled off to be recycled for use by golf courses, sod farms, street mediums, and parks. Some wastewater plants in other parts of the country have gotten so high tech that the water once having gone through the process can be and is safe to drink.

Worldwide Water

In most developed countries, water is supplied to households and industries using underground pipes. That water is processed and treated to meet drinking water standards, even though only a very small portion is consumed or used in food preparation. In the United States, less than 1 percent of municipal water is used for human consumption. The rest is used for things like bathing, water gardens, cleaning and cooking. Municipal drinking water is regulated by the E.P.A.

The quality and reliability of municipal supplied water can vary from community to community. If a consumer has questions or concerns about their municipal supplied tap water, they should review their cities utility's Consumer Confidence Report. The E.P.A. also provides additional information about drinking water that consumers may find helpful. In most cities and towns, municipal water comes from large wells, lakes, rivers, or even reservoirs. Cities and towns process the water at treatment plants like the James W. Jardine water purification plant as discussed earlier.

The water is tested for E.P.A. compliance and is then piped to residential homes and industries.

This is not the same on a global level. India for example has the world's highest number of people without access to clean drinking water, imposing a major financial burden for some of the country's poorest people.

According to the Water Aid, an international charity that strives to improve access to safe water, as well as hygiene and sanitation. Almost 76 million people in India (5 percent of the population) are forced to either spend an average of .72 cents per 50 liters of water a day (20 percent of their daily income) or use supplies that are contaminated with seweage and chemicals. The people of Britain spend around .10 cents per day for the same amount.

Many people are forced to turn to an alternative in order to access water, due to high prices, or the simple issue of accessibility.
Unsanitary water comes with consequences and is the cause of countless illness each year.
There are about 315,000 children who die from diarrheal diseases each year, 140,000 of which occur in India. A major issue in India is the mismanagement of the water sources, poor planning, and execution.

Not only have the pipelines themselves proven to be less than adequate but also so are the sources from which the water is being used. Rivers and groundwater have become increasingly polluted.

Drought and the sheer population of India play a huge part in the thirst for clean drinking water. As of 2016, India's population has hit 2.3 billion people. It is for the reasons we just discussed that organizations such as,

World Plumbing Council (non-profit organization) founded in 1990 (http://www.worldplumbing.org/) who also created World Plumbing Day; launched in 2010 (an international event every March 11th) with a focus of sanitation and the importance of plumbing, as well as attention of the essential roles of plumbers in maintaining public health and safeguarding of the environment.

2 CODEBOOK NAVIGATION

Old VS. New

The new 2014 Illinois State Plumbing Code Book can be a little challenging to someone new in the trade. This chapter will discuss the various sections and related code topics. I have seen this firsthand when I am teaching apprentices. I ask them by the third or so class, to "put" their new codebooks together.

For those of you that do not know the codebook requires some assembly. Now it is not what you may think, by "putting together" I am referring to putting the pages in order. Followed by the appropriate illustrations and so on.

I have made codebook navigation part of my curriculum and found it to be very helpful to the new apprentices. However the codebook that was given to me in 1987 is a far more difficult to navigate through than the new one.

The code old book is an 8-1/2 x 11 containing 276 pages without any tabs to reference a particular section. In 1994, the book had undergone a much-needed update, not to mention a few code changes along the way.

There was a tabbed section for each part, the illustrations had been updated, and became easy to read and understand.
It may have also been because I was entering my seventh year in the trade.

I also bring it; the old book, to the first day of apprentice training classes as well to show the students how far the codebook has evolved. During the four-hour continuing education classes I instruct, I ask, "How many plumbers in the room remember this old book?" While holding it in the air.
It was back in August of 2000 when the Illinois Department of Public Health made a change to the licensing code section of the codebook.

The effective date was August 1st 2000, Section 750.500

The section reads as follows:
Each licensed plumber shall, as a condition of each annual license renewal after the first license renewal, provide proof of completed in one or more courses offered by course sponsors approved by the department.

Therefore, as you can see by the verbiage that could be a little confusing. All that means is before a licensed plumber can renew their license, he or she must show proof of completing a four-hour continuing education class. However if the plumber is newly licensed that person is exempt their first year. The plumbing department must first approve all continuing education classes.

What this means is the sponsor(s) who are going to instruct a continuing education class; must in writing provide the department with a syllabus of what the class will be. In addition, a resume should be included; after all, there are certain criteria that must be met.

The newest copy of the Illinois state codebook is in a green plastic binder and has been since the mid 90's. It is 373 pages in total, with a lot of verbiage cleaned up and, or removed. There are 18 parts to the book with a total of 18 tabs. The pages are assembled in the following order tabs for each section follow after text and illustrations.

2014 Illinois State Codebook

The following list is the subject tab and the corresponding pages.

1. Illinois Plumbing License Law (LL-1) – (LL-35)
2. Illinois License Code (LC-750-1) (LC-32)
3. Table of Contents (TC – 1) (TC – 9)
4. Definitions & General Provisions (A-1 , A-32)
5. Appendix B thru K Illustrations (APP. B pg. 1) (APP. B pg.-27)
6. Plumbing Materials (B-1 , B-2)
7. Joints and Connections (C-1, C-8) App. C pg. 1-3
8. Traps & Cleanouts (D-1, D-5) App. D pg. 1-8
9. Interceptors & Backwater Valves (E-1, E-5) App. E pg. E-1 -6
10. Plumbing Fixtures (F-1, F-19) App. F pg. F-1-6
11. Hangers Anchors & Supports (G-1, G-3) App. G pg. G-1-4
12. Indirect Waste Piping / Special Waste (H-1, H-3) App. H pg. H-1-6
13. Water Supply & Distribution (I-1, I-25) App. I pg. 1-15
14. Drainage System (J-1, J-9) App. J pg. 1-12
15. Vents & Venting (K-1, K-11) App. K pg. 1-33
16. Correctional Facilities (L-1, L-4) No Illustrations
17. Inspections Testing Maintenance & Administration (M-1, M-3) No Illustrations
18. Appendix A (APP. A-1, A-54) Tables

Illinois Plumbing License Law

The Illinois Plumbing License Law begins with section 1. This section is referred to as the declaration of purpose licensing and standards. It really sums up the dangers of improperly installed plumbing. The one thing to point out is that this, as all other codebooks, has been written with new installations in mind, or "new work." One of my favorite parts of this section is the first paragraph.

It has been established by scientific evidence that improper plumbing can result in the introduction of pathogenic organisms into the potable water supply, result in the escape of toxic gases into the environment, and result in potentially lethal disease and epidemic. If history has taught us one thing that is the first paragraph.

During apprentice training, I will have each person read from this first section aloud. The second paragraph continues with explanation as to why plumbing is as stringent in Illinois. Section 2 is defining who is who, such as what is an agent, a person who is going to sponsor an apprentice under one's license. This section covers exactly what is considered plumbing and establishes who by definition is a plumber.

This second section for the most part is all about ground rules, and identifies the players in the game. Section 2.5 continues to clarify what and who is allowed to install irrigation systems, what an irrigation contractor must due in order to satisfy the plumbing code. It also discusses what the department (Illinois Department of Public Health) will and will not do. Now for those of you, that either new to plumbing or just not sure, why irrigation is so important in the plumbing code, it is for the simple reason that irrigation equipment is connected directly to the potable water supply. How is this a bad thing? You ask. Well first off, let us look at some of the pesticides used in the lawn care industry. However, as we do I would like you to re-think paragraph one of the codebook. Contrary to what the lawn care industry would have you believe herbicides and pesticides are far from "safe." They include components of wartime defoliants like Agent Orange, nerve gas type insecticides, and artificial hormones.
Some the Federal Government has prohibited from use on its own property. What does this have to do with plumbing; well two things are going on. First if we have an irrigation system with a direct physical connection to potable water. (In which I will elaborate on in chapter 10.)
Remember water is always fluctuating in pressure, so as pressure drops and increases water will move back and forth in the municipal water main much the same as "sloshing" liquid in a container. As that happens any water and chemicals that are on the lawn, or in the proximity of the sprinkler head could potentially be pulled back into the drinking water by means of creating a vacuum.

This process is referred to as back siphonage. Anytime there is a negative, force or vacuum and reverses the flow of gas, air or water towards the opposite direction of intended. Some herbicides can remain active for years after they have been applied. So if a vacuum or back siphonage occurs the herbicides or pesticides can flow into the potable water.

The only way that this condition can be protected is by installing a backflow preventer on the irrigation service line. How do you know which one to install? There are two types of hazard classification in plumbing, especially when we discuss backflow devices.

We have a high hazard and a low hazard. So which category would this scenario fall under? If you guessed high hazard you would be correct. I have left out another contaminant. Can you figure out what? Animals can contribute to the contamination of water supply as well as man, although man is probably much better at it.
Section 2.6 under the Plumbing License Law covers more irrigation as it applies to what a golf course maintenance staff can and cannot be permitted to do.

Section 3 Begins with persons allowed designing plumbing systems. Now a plumber and apprentice surely can design, install, and alter; however, when the job is complete only the licensed plumber can inspect the work. A licensed architect and a licensed engineer can also design plumbing systems.

This section also spells out owner and tenant type buildings and who can repair plumbing. For example, a homeowner is allowed to make repairs, construct, alter, and install plumbing in their own home; they are however required to get the proper permits and comply with the plumbing code in its entirety. The catch is that the home must be their primary residence for not less than 6 months.

Section 6 of the codebook deals with advertisement. Plumbing and advertisement is straight forward; if you advertise plumbing on anything from business cards to flyer's to website, trucks, or anything else you can stamp your name, you must include a plumbing license number.
Now I have had owners of companies tell me "I don't want my number out there like that" My response is always the same, use your contractor number not your personal license number. As of May 1st 2002 if a plumber wanted to become a plumbing contractor in Illinois they were required to submit an application to the department with the appropriate fees paid and proper licenses, and must be bonded and insured.

They would then be given a unique contractors number. The way it used to be done prior to the change in 2002, the plumbing contractor had to become bonded in the city, village, or town they wished to operate in. The contractor would have to provide insurance, and a copy of his/her plumbing license.

Of course this lead to individuals willing to "pull permits" for the right price. The numbers all begin the same prefix for every plumbing contractor in Illinois it is commonly referred to as "your" 055 number.

Here is a perfect opportunity to bring up that all plumbing licenses begin with a 058 prefix. Much like the 058 assigned to the plumber. The 055 is a company assigned contractor number. Now for you apprentices your license number will begin with a 056. This is how it is quickly determined if you are a licensed plumber, contractor or apprentice.

In addition, section 3 will get into specifics of license holders for companies. I have been a license holder for most all of the companies I have worked for throughout my career. What this means is that I as a licensed plumber understand and accepts the role of being the licensed professional. There is also some degree of risk associated with this responsibility. Early on in my career in fact:

I passed my exam in November and by January; I was a license holder for a shop. By February, I had enrolled in a cross connection device inspector's class.

Now back to becoming a license holder, in Illinois at least one licensed plumber typically the person holding the shop license must be an officer of the corporation.

This could mean (back then) the license holder was required to be a part of the corporate structure as one of the following. Vice president, Treasurer, Secretary and so on.

A retired plumber who no longer is engaged in plumbing but would like to keep his/her plumbing license can do so but they are not eligible to be a license holder for a shop, nor can the inspect plumbing. A retired plumbing license is just that, it is dormant.

Let us look at how an individual becomes a licensed apprentice. Yes, of course there are criteria that one must meet. First off, the applicant must fill out the appropriate forms.

The forms are available on line by going to the Illinois Department of Public Health website and downloading them under the forms tab. The applicant must also be at least 16 years old, and have a licensed plumber willing to sponsor him/her. Once the form has been completely filled out the applicant must mail it to the department with an enclosed check and two 1"x 1" passport photos of the applicant's head and shoulders. No hats or sunglasses should be worn in the photo.

Now as far as a licensed plumber becoming a sponsor for an apprentice, he/she will also need to sign off on the application and provide some basic information. Name, address, and license number, the applicant for the apprenticeship is not authorized to work as an apprentice until all the paperwork is complete, received by the department for approval.

The department will then send a wall certificate and a wallet card to the sponsors (licensed plumber) address provided on the application. The sponsor also has a responsibility to field train, and instruct the apprentice.

Now what I can tell you is that never happens. Even during my entire apprenticeship, not once did I ever attend classroom type of training. This just did not exist back then, everything I learned about plumbing was from the time spent in the field.

I was non-union for most of my career; the unions have training in classroom settings. I was paid the same, had a benefits that included medical, health, eye care, 401K but the one thing we; as non-union plumbers didn't have was the formal training. Now I was a union plumber for about 10 years, I came in the "back door" as I would hear almost daily from my colleagues.

I signed up with the union and paid my way in. I was already a licensed plumber when I started, and I remember feeling a little nervous my first day. Here is why, I was expecting to be working alongside individuals that had formal training, and I knew I did not.

I was really looking forward to learning everything they knew, wow was I wrong! When it came right, down to it, they did not know anymore or any less than I did. On the same token, several of the person's I met and worked alongside of were as focused as I was when it came to codes, installation procedures and so on.

A sponsor can only have two apprentices under his/her license at the same time. The rule for that is the sponsor can have two at the same time as long as the apprentices are 24 months apart. An apprentice also cannot renew his license before his/her sponsor. The apprentice license can be renewed up to a total time of six years.

It is at that point the apprentice either takes the exam or begins a new career. The department once again has the exam forms application available on line. Section 3 wraps up with the description of private water systems and municipality systems.

The main difference is that a private water source is for a single building such as a well, whereas a municipal water supply is public use. The municipal water supply and the individuals who operate them on a daily basis are not required to hold a plumbing license. The same holds true for a private system. There are a few other topics in this section of great interest, but for the purpose of this book, we will move on to section six.

This is a short section, but worthy of bringing up only because it reminds me of probably one of the funniest plumbing inspector related stories I have ever heard. Section 6 reads.

No municipal corporation or political subdivision shall engage in plumbing unless one performs such plumbing or more licensed plumbers, or licensed apprentice plumbers under supervision in accordance with this act, if any such governmental unit may contract for plumbing with any person authorized to engage in plumbing in this state.
First, let us take a look at what is a municipal corporation is a town, city, and village. Of course, a corporation is an entity capable to conducting business. The story I have heard, and I must tell you this is a rumor, as I cannot confirm this. A village in the suburbs had a water main break near the village hall. The water department workers had excavated a hole; shut the water off so that the repair could take place.

A water main is usually made of ductile iron pipe that is a very heavy and thick walled pipe. The repair method is typically the same on most of these type of breaks. The repair can then be accomplished by using what is known in the industry as a water main sleeve; this is nothing more than a neoprene interior with a stainless steel wrap around it.

There is also a "claw" type of clamp with a series of bolts that once wrapped and tightened around the pipe it will seal the cracked pipe. The plumbing inspector had stopped by to check on the progress of the water department.

Upon further investigation, the plumbing inspector discovered that none of the water department workers had a plumbing license. The plumbing inspector, mind you an employee of the very same village as the water department, fined the same village he worked for under section 6 of the codebook.

Now again I cannot confirm this to be true but whenever this story comes up, I laugh every time I hear it. This individual by the way had turned me into the state more than once as he "caught" me working as an apprentice without direct supervision.

I am very familiar with this person and knowing him, I lean towards this story being true. What is direct supervision?

Direct supervision is an apprentice with less than two years on an apprentice license. The apprentice may install and repair plumbing, but he/she is required by state license law to be supervised by a licensed plumber. This is for the first two years of the apprenticeship.

At the third year, the direct supervision will then be removed from the license, and is replaced with "supervision." Supervision will be clearly printed on the apprentice's license as is direct supervision is. This will remain on the license until the apprentice becomes a licensed plumber. In Illinois, an apprentice is not eligible to take the exam, until they have served out 48 months.

With that, apprentices are also required to work a minimum of 1500 hours per year. Now for some insight on the actual exam, the first thing we will look at is the exam process and the people that are instrumental in the process. An Illinois State Board of Examiners made up of nine licensed plumbers designated from time to time by the director.

The board serves to aid the director and the department in a variety of ways. The board prepares the subject matter for the examinations; suggests rules governing examinations, as well as hearings for suspension revocation or reinstatement of licenses. Submitting recommendations to the director makes for an efficient administration of this act.

The grading of all exams for licenses, and performing other duties as prescribed by the director. The job of director includes such duties as the preparation of forms for the application for the exam, prepare, and issue licenses.

With the help of the board, prescribe rules and regulations for examination of the applicants for plumbers' license. Other duties are the issuance of plumbers' license, revoke, suspend, and renew.

The director also is responsible to maintain records of registered plumbing contractors, licensed plumbers, apprentice plumbers, and retired plumbers.

This also include the issuing dates, hearing for denial, suspension, or revocation of any license. The director with the board will also define what constitutes as an approved course of instruction in plumbing. The department will hold examinations for applicants for plumbers' licenses at least once every three months.

That has not been the case recently with the budget problems in the state. The budget has effected all types of occupations, including the plumbing program. Illinois has been operating without a budget in place for about 14-18 months at this point. With a recent phone call to the plumbing program, I had been told that the state has run out of supplies to issue licenses. Illinois cannot get ink, laminate supplies, or postage in which to create the license and mail them.

The regulations regarding the exam are straightforward. An applicant must file the written application with the department at least 30 days before the date of the exam. The applicant must pay the exam fee, submit proof of citizenship, has completed at least two years of high school, and has worked as an apprentice plumber for at least four years.

Once the applicant has been approved to take the exam, the department will notify the applicant immediately, identify the time, and place that the exam will be held. If an applicant fails, the exam may reapply by filling out the correct forms for a retake of the exam. This brings us all the up to section 14, and license renewal. Assuming you have passed the exam now, you are required to obtain not less than four hours of continuing education.

This is the point in your career that our paths may cross. I teach these classes consisting of code review as well as any new code topics that have been added, altered, or changed. The department will not allow you to renew you are license without the four continuing education course being completed.

How does one become an instructor as a sponsor registered with the department? You got it! Fill out the appropriate form; submit in writing your intended syllabus for approval. If you are not a certified instructor, you must be a licensed plumber. If you should fail in the renewal process for five years the department will toss all of your records, and you have to retake the exam again. The same is true for a licensed apprentice. However, a licensed apprentice is not required to submit the four hours of continuing education. In my opinion, it is a great idea for apprentices to participate in the continuing education classes. I always welcome licensed apprentices to participate in my classes. Even though some of the information maybe over their heads, it is still a good experience for them.

This brings us to section 16, which states any town city or village with a population of 5000,000 or more can write, and issue plumbing codes and licenses. The city of Chicago is the only other licensing agency in the state. They may also provide a plumbing board of examiners administer exams. They also issue plumbing licenses, apprentice licenses and plumbing contractor licenses.

Chicago has its own plumbing code that is similar yet different from the state plumbing code. What this means if you work in the City of Chicago make sure you understand there is a big difference in codes; we will look at some of the differences in another chapter. Depending on who you ask about which code is better, or more stringent one thing is for certain, you'll definitely get an opinion way. As I mentioned earlier, I am a licensed plumber through the Illinois Department of Public Health.

I also have a Chicago contractor's license issued from the City of Chicago. I did not however have to take an exam for the later, my state issued licenses allows me to work anywhere in the state including Chicago. The same is true for people that hold a Chicago journeyman's license. An individual that holds a current Chicago license can also work anywhere in the state.

The City of Chicago has a population greater than 500,000 as I said before they can write their own plumbing codes. On the registration side, I must be registered as a contractor, which of course equals revenue.
A Chicago licensed plumber must also be registered with the state if he/she is going to become a plumbing contractor.
On both sides Chicago and state there is a fee as expected. In section 18-29 of the license law covering various topics such as suspension, of licenses hearings, licensing from another state, which by the way an individual with a plumbing license from another state will be required to successfully pass the Illinois exam before obtaining a plumbing license in Illinois.

I would like to point out just a few fines that one could face if that person should attempt to practice or holds himself or herself as a plumber, or plumbing contractor without the proper licensing. The maximum fine is $5,000.00 for each offense, and a hearing to determine the civil penalty.
As you can see this can be quite a financial burden if you are caught doing this.

There are other things to consider before engaging in or attempting to work as a plumber or contracting work as a plumber.
The steep fines aside, if there are violations to the plumbing code not only can you be fined but also financially responsible to pay a licensed and registered contractor to make all code violation corrections.

Permitting and Inspections

Let us assume you are licensed as a plumbing contractor and have a customer who has hired you to change their water heater. What is next? Its time to get a permit from the village, town, or city in which you will be doing the installation. Here are a few tips of what is required before the permit can or will be issued.

You will need a letter of intent, which is a state plumbing license law under section 37. A letter of intent is on company letterhead including contractor's license number, plumbing license number, and the customer's name address. In the letter you will describe what your intentions are, I have found in writing these over the years simple is best, don't over think it, be as direct as possible. Some villages, towns, or cities may require you to submit a drawing of the alterations if it is a remodeling project.

For this example, let us stick with our water heater project. There may or may not be a permit fee required by the village, I know for repairs to the sanitary sewer for an example in some towns there is no permit fee.

Your letter of intent should also include your corporate seal or have it notarized once again under section 39 of the license law.
The village will also require a permit application to be filled out.

Typically, this will contain basic information property address, owners name and the type of work to be performed. You should also have a copy of your plumbing license and a copy of you state registration or contractor's license.

I keep both on one page for ease and less paperwork. For the most part plumbing permits depending on the size of the project can be obtained immediately over the counter. The larger projects may require a plan review; this will most definitely slow the process.

While at the counter, you may want to consider setting up a plumbing inspection. A water heater install should be a quick job of just an hour or two. In most cases, the plumber is not required to be present during such an inspection. It is professional curtesy however so I suggest and try to be available anytime you have a scheduled inspection.

A good relationship with the local plumbing inspector is not a bad thing to have. Remember all plumbing inspectors have worked in the field at one time or another during their career and some still do. If you do a lot of work in one or two areas, it would be beneficial to be able to have a good rapport with the inspector.
While working as a plumbing inspector one of the most questions I would get was "what type of plumbing jobs require a permit." My answer was "a permit is required for plumbing," and the follow up question was "okay but what type of plumbing?" Once again, I would reply "any plumbing work." In addition, not to be a wise guy I would suggest looking up the definition of plumbing in the codebook. It can be found in the license law section under "plumbing."

Plumbing is defined as all piping, fixtures, appurtenances, and appliance for the supply of water for all purposes. This includes without limitation lawn sprinkler systems and backflow prevention devices connected to lawn sprinkler systems, from the source of a private water supply on the premises or from the main in the street, alley, at the curb to, within, and about any building or buildings where a person or persons live, work, or assemble.

Plumbing includes all piping, from discharge of pumping units to and including pressure tanks in water supply systems.

An easy thing to remember when discussing plumbing is the domestic water system in its entirety is considered plumbing. Now when discussing the waste or building drain, sewer things change slightly.
The sanitary for a building is considered to be terminated under the definition of plumbing, once it is beyond five feet of the foundation of the building. At this point, the sanitary sewer falls under the drain layers license. However, that "rule" is subject for debate in different municipalities. It would be best practice to check with the building department for clarification on this.

Some municipalities are now requiring that a licensed plumber install, and repair the pipe. For you sewer and water contractors this could mean you will be required to either obtain a "055" plumbing contractor's license or at the very least hire a licensed plumber.

In addition, do not forget he will need to be listed as a member of the corporation. One last thing I will add to this section is do not be surprised if the inspector require you to perform a stack test on you work. This means under 890.1930 of the codebook an inspector can ask for a stack test of the waste and vent system.

This is more often completed during the "rough in" stage of new the construction of a building. The reason is simple the plumbing inspector is checking to be sure that the system is gas and watertight prior to insulation and drywall. It is also common for this to be requested on a repair job also.

The gas and water tight that I am referring to can be found in the codebook and is under 890.310 joints and connections tab. If you are wondering just what is a stack test? In simple terms, it is nothing more than filling the waste and vent system to a predetermined height, usually ten feet of head pressure.

Which is the recommended head pressure by most if not all PVC pipe manufactures. Let us not forget about pressurizing the domestic water system also. This can be achieved in two ways one simply turning the water on, or two by means of an air test. I would recommend the latter if you are testing an unheated building and the outdoor temperature is below 32 degrees.

3 PLUMBING DEFINITIONS AND TERMS

As with any other industry there is widely used terms, or slang in which plumbing fittings and parts are referred. In this chapter, we will cover the proper definitions and terms used in plumbing today. This is something I discuss in my apprentice training classes.

I explain to the apprentices that communication within this industry is of high importance. If you speak "plumber jargon" when explaining problems to potential customers you will most certainly have them scratching their heads. (See chapter 14) The same is true if you are at the supply house counter and you are asking for a double flam shooter, "homeowner jargon" for a problem with a faucet.

With both scenarios, there is some middle ground in which to communicate effectively with a potential buyer of your services, and others within the trade. I have included in the following a list; complete with the definitions of each of the most common terms used within the industry.

Much of the terms listed are used throughout this book. For further information regarding the terms and definitions used, they can be found in your codebook in section 890.120 under the tab Definitions & General Provisions. For the purpose of this book, I will be sticking to the basics.

Definitions

Accessible: Easily approached or entered with minor modification. This includes the removal of an access panel, drywall, paneling or similar. Concrete, ceramic tile and asphalt are not accessible by this definition.

Air gap: The unobstructed vertical distance through the free flowing atmosphere, from any pipe or faucet supplying water to a tank or plumbing fixture and the flood level rim of the receptacle.

Antimicrobial: An additive or surface coating that will prohibit the growth of bacteria.

Approved: Accepted under an applicable specification stated or cited under the plumbing codes as per the department and Illinois Plumbing Code.

Aspirator: A device supplied with water under positive pressure that passes through an integral orifice; causing partial vacuum which results in fluid moving by way of siphonage.

Atmospheric vacuum breaker: A device consisting of mechanical moving parts used for the protection against back siphonage.

Authorities having jurisdiction: Any entity that the Illinois Plumbing License Law authorizes to enforce the Law.

Backpressure: A condition caused when a force is exerted and reverses the flow of water, gas, or air in the opposite direction in which was intended.

Back siphonage: A condition caused when a negative force or vacuum is exerted reversing the intended direction of flow.

Backwater valve: A device or valve install in a sanitary sewer, or storm line to prevent water backing up into a building.

Backflow: The reversal of flow from a direction of originally intended. This includes backpressure and back siphonage.

Backflow preventer: A device or an assembly used to prevent the contamination of potable water.

Battery of fixtures: Any group of two of more identical fixtures that discharge into a common horizontal waste branch.

Boiler blow down: A controlled outlet on a boiler to allow emptying of sediment.

Branch: Any part of the piping system other than the main, riser, or stack.

Branch vent: A horizontal vent connecting one or more individual vents with a vent stack.

Building drain: the lowest part of the horizontal drainage piping, within the walls of a building. The buildings drain terminates five feet outside the buildings foundation.

Building sewer: The portion of horizontal piping that continues from the building drain five feet beyond the foundation.

Building storm drain: The lowest horizontal portion of the drain, which receives water from rain, ground water, surface water, or clear water waste such as condensation piping.

Building sub-drain: The portion of sanitary drainage system that cannot be drained by gravity.

Chemical waste: Piping system that conveys such waste as toxic, or corrosive to the drainage system.

Circuit vent: A branch vent that serves two or more traps extending from the front of the last fixture connection of a horizontal waste branch to the vent stack.
(Only applied to floor drains and floor outlet fixtures.)

Clear water waste: Cooling water and condensation waste, refrigeration, or air conditioning equipment.

Closed water system: A system that has a backflow device install the water supply, to contain backflow within a building.

Code: State or local statutes, ordinances, or administration rules, and requirements for plumbing methods, or materials.

Cold water: Water below 85 degrees Fahrenheit.

Combination fixture: A fixture combining two or more compartments, or receptors.

Combination waste and vent: A system of waste piping with the horizontal wet venting of one or more floor drains by means of a common waste and vent adequately sized to provide free movement of air above the flow line of the drain.

Combined building sewer: A sewer that receives both storm and sanitary waste.

Common vent: A vent connecting at the junction of two fixture drains and serving as a vent for both fixtures.

Connection: The joining of two pieces of pipe, or valves, fittings and other appurtenances.

Contamination: Any solid, liquid, or gaseous matter that, when present in a potable water supply or distribution system may cause water to degrade so that the quality of the water may cause illness injury or be fatal if consumed.

Contaminated water: Water not suitable for human use or that does not meet the water quality standards of rules of the Illinois Pollution Control Board titled Primary Drinking Water Standards.

Cross-connection: Any actual or potential connection or arrangement between two otherwise separate piping systems, one containing potable water and the other containing any kind substance not suitable for human consumption.

Cross-connection control device: A plumbing appurtenance installed in a potable water line to prevent any substance of any kind from being mixed.

Dead End: A pipe that is terminated at a developed distance of two feet or more by means of a plug or other closed fitting, except piping being served as a clean out.

Department: Illinois Department of Public Health.

Developed length: The length of a pipe measured along the centerline of the pipe including fittings.

Diameter: The length of a straight line passing through the center of an object, or a circle.

Drain: Any pipe that carries wastewater in a building drainage system.

Drain laying: The laying and connecting of piping from five feet outside the foundation wall of a building to the public sanitary sewer system in the street or alley.

Drainage fixture unit: The mathematical factor used by the plumbing industry to establish the estimated probable load on the drainage system. One drainage fixture unit is equivalent to 7.5 gallons per minute or one cubic foot per minute.

Durham system: A soil or waste system where all piping is of threaded pipe, using recessed drainage fittings.

Existing plumbing: A plumbing system or any part of a plumbing system that has been installed prior to January 1, 2014.

Fixture branch: A water supply, soil pipe, or waste pipe serving one or more fixtures.

Fixture carrier: A device designed to support an off the floor plumbing fixture.

Fixture drain: The vertical or horizontal outlet pipe from the trap of the fixture to the junction of that pipe with any other drainpipe.

Fixture supply: A water supply pipe connecting the fixture to a branch or main water supply pipe.

Fixture supply stop: A valve used to control water supply to an individual plumbing fixture, appurtenance, or appliance.

Flood level rim: The top edge of a receptacle or fixture over which a liquid will flow when the receptacle is filled beyond its capacity.

Flush valve: A device for the purpose of flushing water closets and urinals.

Grade: The fall, pitch or slope of a line of pipe in reference to a horizontal plane. In drainage, it usually expressed as the fraction of an inch fall per foot length of pipe.

Graywater: Untreated wastewater that has not encountered toilet waste, kitchen sink waste, dishwasher waste or similarly contaminated sources.

Grease interceptor: A device used to separate and retain grease, oils, and other floating matter from seweage waste while permitting the remaining flow to discharge into the drainage system.

Group of fixtures: Tow or more fixtures adjacent to or near each other.

High hazard substance: Any substance that, when present in the potable water supply system can cause illness, injury or be fatal if consumed.

Horizontal branch: A drainpipe extending laterally from a soil or waste stack or building drain, with or without vertical sections or branches, which receives the discharge from one or more fixture drains and conducts the discharge to the soil or waste stack or to the building drain.

Horizontal pipe: Any pipe of fitting that makes an angle of less than 45 degrees with the horizontal.

Hose bibb: A faucet to which a hose maybe connected.

Hot water: water temperature of not less than 120 degrees Fahrenheit.

Indirect waste: A pipe that does not connect directly to the drainage system but conveys liquid waste by discharging through an air gap into the drainage system.

Individual dry vent: A pipe installed to vent a single fixture trap that connects with the vent system above the fixture served, or that terminates in the outside atmosphere.

Interceptor: A device designed and installed to separate and retain hazardous matter from normal waste and to permit normal seweage or liquid waste to discharge into the drainage system. Interceptors may be designed to remove gas, oil, sand, and grease.

Invert: the lowest part of the internal cross-section of a pipe.

Island fixture vent: A vent in which the vent pipe rises as near as possible to or above the highest water level in the fixture vented and the turns down before rising to connect to the vent system 6 inches above the flood level rim or terminating to the atmosphere.

Joint: The juncture of two pipes, a pipe and a fitting, or two fittings.

Lead free: When used with respect to solder and flux refers to products containing not more than 0.2 percent lead. When in reference to wetted surfaces of pipe, fittings, or fixtures not more than 0.25 percent of lead is allowed.

Length of pipe: The overall distance measured along the centerline of a pipe.

Line valve: a valve in the water supply distribution system, except those immediately controlling one fixture supply.

Load factor: The percentage of the total connected fixture unit flow rate that is likely to occur at any point in the drainage system. The load factor varies with the type of occupancy, the total flow above the point being considered, and probability of simultaneous use. Load factor represents the ratio of the probable load to the potential load.

Loop vent: A circuit vent that loops back to connect with a stack vent instead of a vent stack. Its use is limited to floor drains and floor outlet fixtures.

Low hazard substance: Any substance that, when present in the potable water system, may cause the water to be discolored or have an unusual odor or an unpleasant taste, but will not cause illness, injury, or death if consumed.

Main: The principal artery of a piping system to which branches may be connected.

Main vent: The principle artery of the venting system to which branch vents are connected.

Maximum demand: The greatest requirement of flow of either water supply or waste discharge.

Metering faucet: A self-closing faucet that dispenses a specific volume of water for each actuation cycle.

Minor repairs: Repairs that do not require changes in the piping to or from plumbing fixtures or involving the removal, replacement, installation, or reinstallation of any pipe or plumbing fixture.

New plumbing: Any plumbing system or part of a plumbing system or part of a plumbing system, or any addition to or alteration of any existing system, being installed or recently completed.

Non-potable water: Water that does not meet drinking water quality standards.

Offset: A combination of elbows or bends that bring one section of pipe into a line parallel with another section.

Overflow rim: The top edge of a receptacle or fixture over which a liquid will flow when the receptacle or fixture is filled beyond its capacity. (Interchangeable with flood level)

Peppermint oil: A pungent, aromatic oil sometimes used in testing a drain waste system by means of a peppermint test.

Peppermint test: A test for leakage using peppermint oil and hot water to determine any leak in the drainage and vent system.

Pipe increments: Increasing or decreasing pipe size by given number as in the following examples: 1/2, 3/4, 1, 1-1/4, 1-1/2

Plumbing appliance: A special class of plumbing fixture intended to perform a specialized function. Including water heaters, water coolers, drinking fountains, and water treatment equipment other than water softening equipment.

Plumbing appurtenance: An accessory or device used in a plumbing system, which demands no additional water supply, or adds any discharge load to a fixture or the drainage system. Plumbing appurtenance includes instruments, gauges, backflow assemblies, relief valve, limit switches, and solenoid valves.

Plumbing fixture: Approved, installed receptacles, devices or appliances that are supplied with water or that receive or discharge liquid or liquid-borne waste, with or without discharge of the waste into the drainage system to which they may be directly or indirectly connected. An installed appurtenance to the potable water supply system that makes available intended potable water, or receptor that receives and discharges liquids or liquid-borne waste either directly or indirectly into the drainage system; or a permanent appendage usually designed as a receptacle and intended to receive or discharge liquid or liquid-borne waste to a drainage system. Industrial or commercial tanks, vats, and similar processing equipment are not plumbing fixtures, but they may be connected to, or discharged into, approved traps or plumbing fixtures.

Plumbing inspector: An employee or agent of the state or local government who holds a valid Illinois state plumbing license and is authorized to inspect plumbing.

Potable water: Water that meets drinking water quality standards specified in the Pollution Control Board's rules titled Primary Drinking Water Standards and is suitable for human consumption or culinary use.

P.S.I.: Pounds per square inch

Public area: An area within a building accessible to all persons, including, but not limited to, mercantile units, private clubs and membership organizations.

Ready accessible: Direct access without the necessity of removing or moving any panel, door, or similar obstruction.

Receptor: Devices or fixtures that receive the discharge from indirect waste pipes.

Relief valve:
1. Temperature and relief valve – A valve designed to release water to the atmosphere at a predetermined temperature setting.
2. Pressure relief valve – A valve designed to relieve excessive pressure to the atmosphere at a predetermined setting.
3. Temperature and relief valve – A valve incorporating a temperature pressure relief valve in one unit.
4. Vacuum relief valve – A valve that admits air to the system when the system is attempting to reduce its pressure to less than atmospheric.

Return offset: A double offset installed to return the pipe to its original alignment.

Riser: A water supply that extends vertically one full story or more to convey water to branches or to a group of fixtures.

Rough in: The installation of all parts of the plumbing system that can be completed prior to the installation of fixtures. This includes drainage, water supply, and venting piping, and the necessary fixture supports.

Safe pan: An appurtenance installed beneath piping or a fixture to collect and drain leakage.

Service connection: The tap at the water main and any piping to the property line.

Sewage ejector: A device for lifting sewage by pumping means,

Soil pipe: Any pipe that conveys the discharge of water closets or fixtures having similar functions, with or without the discharge from other fixtures, to the building drain.

Stack: Any vertical line of soil, waste or venting.
Storm sewer: A sewer that is used for conveying rainwater, surface water, ground water, site drainage, condensate, cooling water or other similar liquid. Excluding sewage.

Sub-soil drain: A drain that collects sub-soil water and conveys it to a place of disposal.

Sump pump: A pump used for the removal of storm, subsoil, and clear water waste drainage from a sump.

Tempered water: Water ranging in temperature from 85 degrees to but not limited to 120 degrees in Fahrenheit.

Trap: A fitting or device designed and constructed to provide, when properly vented, a liquid seal that will prevent the back passage of air without materially affecting the flow of sewage or wastewater through it.

Trap arm: The portion of a fixture drain between a trap and its vent.

Trap primer: A device or system of piping to maintain a water seal in a trap.

Trap seal: The vertical distance between the crown weir and the top of the dip of the trap.

Tuberculation: A condition that develops on the interior of pipe due to corrosion, resulting in the creating of small, hemispherical lumps on the inner walls of the pipe.

Union: A coupling device used to join two pipes end to end, but allow them to be disconnected and re-connected; this joint can be assembled and disassembled without removing any adjacent pipes.

Vacuum: A pressure less than atmospheric pressure sometimes referred to as suction. It is usually measured in inches of mercury below atmospheric pressure, such as 10 or 20 inches of mercury. To vacuum also means to siphon.

Vent pipe: A pipe in a plumbing system that is used to equalize pressure and ventilate the plumbing system.

Vertical pipe: Ant pipe or fitting that makes an angle of 45 degrees or less with the vertical.

Water closet: A fixture with a water-containing receptor that receives liquid and solid body waste and on actuation conveys the water through an exposed integral trap into the drainage system. (Toilet)

Water distribution pipe: A pipe within a building that conveys water from the point of the service line to the point of use.

Water hammer: A concussion or sound of concussion of moving water against the sides of containing pipe or vessel due to a sudden stoppage of flow. Also referred to as hydraulic shock.

Water hammer arrester: A device used to absorb hydraulic shock.

Water heater: A plumbing appliance used for heating domestic water or commercial purposes.

Water main: A water supply pipe for public or community use.

Water supply fixture unit (WFSU): The mathematical factor used by the plumbing industry to estimate the probable demand on the water supply system.

Wet vent: A vent that also serves as a drain.

4 PLUMBING MATERIALS

Drain Waste and Vent

When it comes to choosing materials for any particular job, the first thing to consider is what has been approved on the local level. We are first going to start out with the approved materials that are suitable for the drainage system. Moreover, by approved; I am referring to materials that the state (Illinois Department of Public Health) has listed as approved. As with any job, it is extremely important to make the necessary calls on a local level to determine if that village, town, or city has any addendums to the state code.

Back in March of 2014, I had attended a four-hour continuing education class that members of the department had been speaking at. One topic that was brought up was that going forward; there would be no addendums to the state plumbing code. That has changed and the department has begun accepting code addendums.

The other interesting topic that was also talked about was that anyone who is inspecting plumbing anywhere in the state of Illinois must be certified to do so. That also has since been changed back to anyone with a plumbing license can inspect plumbing work. With that, we are back to allowing villages, towns, and cities to make an addendum to the code. Before we discuss what types of materials can or cannot be used, let us look at the agencies that are acceptable, and who test and stamp their approval on the materials to be used in the plumbing system.

When referencing sewer or waste pipe, the following agencies will have one of the following stamps located on the piping and fittings. In addition, it is important to note that- per Illinois State Plumbing Code 890.210- all materials piping, fittings, appliances, appurtenances, faucets, fixture fittings, fixtures, and devices used in all plumbing systems shall be approved by the department, in accordance with the following criteria:
1. Compliance with the requirements of this part.
2. Compliance with the applicable standard

3. Labeled by an agency that is approved by the department or is an ANSI accredited certification program.

Labeling indicates that the agency certifies the plumbing material to be in compliance with applicable standards. Labeling also includes the manufacture's identification of material. Each length of pipe, each pipe fitting, trap, fixture, device and appurtenance used in a plumbing system shall have cast, stamped, or indelibly marked on it the maker's mark or name, the weight, type, class of product and the standard that applies.

ANSI	American National Standards Institute
ASME	American Society of Mechanical Engineers
ASPE	American Society of Plumbing Engineers
ASSE	American Society of Sanitary Engineers
ASTM	American Society for Testing and Materials
AWWA	American Water Works Association
CISPI	Cast Iron Soil Pipe Institute
CSA	Canadian Standards Association
FM	Factory Mutual Approvals
NSF	National Sanitation Foundation
PDI	Plumbing and Drainage Institute
UL	Underwriters Laboratories

The approved materials listed below are approved by the department for use in building drainage and vent piping.

(ABS) Acrylonitrile Butadiene Styrene Brass pipe
Cast iron

(CPVC) Chlorinated Polyvinyl Chloride

Copper pipe Type L, K, M, and DWV (see footnote 2) Galvanized pipe (see footnote 2)
Glass Fiber Borosilicate pipe (see footnote 3)

High Silicon Content Cast Iron pipe (see footnote 3) Polypropylene pipe (see footnote 3)
(PVC) Polyvinyl Chloride pipe and fittings

(PVC) Polyvinyl Chloride with cellular core (see footnote 4) Polyvinylidene fluoride (see footnote3)
Stainless steel type 304 and 316L Stainless steel butt-weld fittings Stainless steel flanges

Footnotes as follows:
1. Solvent cement must be handled in accordance with ASTM 402-1993
2. Type M copper tubing, DWV copper tubing, and galvanized steel pipe are approved for above ground use only.
3. Approved for corrosive waste or corrosive soil conditions.
PVC pipe with cellular core is approved only for gravity drainage and venting ASME B.1.20- 1983

The following materials are approved by the department for use in the Building sewer.

Acrylonitrile butadiene styrene (ABS pipe) joints to be solvent cement (see footnote 1)

Asbestos cement pipe

Cast iron soil pipe/fittings including joints made of rubber gaskets, and hubless soil pipe

Copper/copper alloy tubing

Concrete pipe

High-density polyethylene (HDPE) pipe joints to be made of solvent cement (see footnote 1)

Polyvinyl chloride (PVC) pipe joint to be made of solvent cement (see footnote 1)

Polyvinyl chloride (PVC) pipe with cellular core (see footnote 2) joints to be made of solvent cement (see footnote 1)

Vitrified clay pipe (see footnote 2)

Polypropylene pipe (see footnote 2)

Footnotes as follows:
1. Solvent cement must be handled in accordance with ASTM F 402-1988.
2. PVC pipe with cellular core and vitrified clay pipe are approved only for gravity drainage.
3. Dimension ratio (DR) 17 or less
4. Dimension ratio (DR) 13.5 or less

Plastic Pipe

Plastic has revolutionized the plumbing industry. It was introduced in the plumbing trade in 1966. Plastic is of the newest materials in plumbing and has changed the way plumbers work. It is widely used today in all forms of weight classifications and approved uses.

Plumbers must be familiar with several different types and sizes, each with its own properties. Each type of plastic is designed for different applications and has different installation requirements. A pipe's application determines the type of fittings and joints that are compatible. Specific fittings are used for water supply, drain, waste, and vent (DWV) systems. Due to the flexibility of the plastic piping, supports cannot be installed at the same distances as cast iron or galvanized materials.

Plastic pipe and materials have advantages and disadvantages. Most plastics are chemically inert, meaning most household chemicals can safely be discharged through them. In cast iron pipe its worst enemy is sugar, plastic can withstand allot of abuse with the exception of high temperatures.

Another advantage is its lightweight, easy to handle, and rapidly put together. The joints are most often solvent cemented together rather than molten lead as with cast iron installations. All plastics are, however, effected by UV rays to some degree. This of course is based on a long-term storage of the materials.

Cold weather will also make plastic brittle; care should be taken when installing plastic in extreme cold conditions, as the cure time of the joints are greatly affected by temperature.

The construction of plastic pipe can vary as well. Pipe can be manufactured with either a solid wall or a cellular core wall. The difference in the two materials is straightforward. The solid wall pipe does not contain trapped air. A pipe of cellular core construction, also called foam core, has walls that contain trapped air. This means it is even lighter in weight than solid wall pipe. Cell core pipe has three layers of plastic- solid inner and outer layers, and a foam middle layer. It is also less expensive than solid core pipe. Plastic pipe is manufactured with various wall thicknesses, commonly called schedules or *size dimension ratios* (SDR)

SDR relates to wall thickness to the diameter of the pipe, using a set ratio of wall thickness to diameter. As the pipe diameter gets larger, the wall thickness increases. An SDR number, such as SDR 35, designates this ratio. The ratio between the outside diameter (OD), or the distance between the outer walls of a pipe, and its wall thickness is constant for each pipe size. The larger the SDR number, the thinner the walls versus the outside diameter.

In an effort to simplify and standardize the use of plastic pipe and fittings, manufactures designated two SDR ratios as schedules 40 (thin wall) which is most commonly used, and schedule 80 (thick wall).

Types of Plastic Pipe and Fittings

ABS, or acrylonitrile butadiene styrene pipe and fittings, are made from thermoplastic resin. ABS is the standard material for many types of DWV systems. ABS is available in diameters ranging from 1-1/2 inches to 6 inches. ABS pipe available in solid and cell core construction. It is also highly resistant to household chemicals. In tests, it showed no effect from common products such as bleach, detergent, and drain cleaners.

PVC, or polyvinyl chloride pipe and fittings, is manufactured from thermoplastic material. The material has an indefinite life span under most conditions. Solid core pipe can be used for pressure systems, but only to carry low temperature water, usually any water above 140 degrees is the maximum that PVC is rated. It also needs to be protected from ultra violet rays, as this will cause it to degrade the thermoplastic materials. PVC is ideal for use in building sewers because it is flexible and will move with ground movement due to climate changes. It is also virtually impenetrable by tree roots.

CPVC, or chlorinated polyvinyl chloride pipe and fittings, are made from engineered vinyl polymer. CPVC is produced in standard pipe size from ½ inch to 2 inches, as are the fittings. CPVC is accepted in many plumbing codes including Illinois. It is manly used in the water distribution system. Its smooth interior, or friction free surface, results in low-pressure loss and higher flow rates. It also provides less chance of bacteria growth. Its molecular structure almost eliminates condensation in the summer months and heat loss in the winter.
CPVC will not rust pit or scale, it is lightweight and installs very fast.

PE, or polyethylene is a thermoset plastic, is flexible and chemical resistant, and its corrosion resistance makes it ideal for the use in transporting chemical compounds. It is most commonly used for the underground gas pipe systems. PE is also inert which means it is unlikely to react with any other substance and is ideal for use in hospitals, beer lines, and soda fountains.
PEX, also known as cross-linked polyethylene, is formed when high- density polyethylene is subjected to heat and high pressure. Because it resist high temperatures, pressure, and chemicals, it is ideal for use in the potable water supply.

PEX is also commonly used in hydronic systems, such as snowmelt, radiant heat, and other applications. However, if it has to be used as potable water it cannot be installed as copper tubing. Inner diameter is not the same as copper tubing. It is smaller; therefore, if you are going to install PEX where you would normally run a ½-inch copper tube, PEX will need to be up-sized to ¾-inch pipe.

PEX is highly flexible and requires the installation of an expansion loop. The same is true when installing CPVC or any other plastics in a potable water supply. An expansion loop is a loop within the piping to allow for expansion. As water moves through the tubing the piping itself will expand and contract, the loop will serve as a shock absorber.

Applications for the plastics listed are as follows:

PIPE TYPE	SYSTEM USE/ APPLICATION
ABS	DWV, sanitary, corrosive waste
PVC	DWV, sanitary systems
CPVC	Water distribution
PE	Gas service, corrosive waste
PEX	Water distribution

In 1968, Thomas Engel invented a process for producing chemically cross-linked polyethylene (PEX) tubing. Considered impossible by many heating industry experts, Wirsbo Co. used Engle's technology to develop a practical manufacturing process for PEX tubing. The cross-linked tubing was introduced to the European floor heating market in 1972 and potable water market in 1973.

PEX tubing solved a number of problems that occurred with metal pipes and some other types of plastic tubing. PEX will not corrode or erode, and is immune to the many problems associated with poor water quality that can damage metal pipes. The tubing is rated at 180 degrees Fahrenheit, 100 psi

The growth of cross-linked polyethylene tubing has been dramatic in the European market. Today, about 10 percent of all plumbing installations are made with PEX tubing and in some countries over 50 percent. In Europe, where radiant floor heating is installed in over 50 percent of all new construction, PEX tubing goes into approximately 70 percent of all jobs.

Tomas Lenman introduced Wirsbo-PEX to the U.S. plumbing market in 1985. Now director of technology at Wirsbo Co. in Apple Valley, Minn., Lenman was a leading member of the engineering team that developed the product in the early 1970s. Today, PEX tubing products are used for floor heating and other heating applications.

Potable Water Materials

As with drain, waste, and vent piping, water supply and distribution piping must also be labeled, remember ALL plumbing pipe, fittings, fixtures, etc. must be approved by the department for use and must also be stamped by one of the agencies as listed previously.

Below is the materials approved by the department for use in water supply and distribution. Some have already been listed as approved in the plastic section listed previously.

Acrylonitrile butadiene styrene (ABS) pipe (see footnote 2) Brass pipe (see footnote 2)
Cast iron (ductile iron) (see footnote 2)
Chlorinated polyvinyl chloride (CPVC) pipe (see footnote 2) Copper/copper alloy pipe (see footnote 2, 3)
Galvanized steel pipe (see footnote 2)
Poly butylene (PB) pipe/tubing (see footnote 2) Polyethylene (PE) (see footnote 2) Polypropylene (see footnote 2)
Polyvinyl chloride (PVC) (see footnote 2) Stainless steel pipe (see footnote 2)
Welded copper tube (see footnote 2)

Footnotes:
1. Solvent weld cement must be handled in accordance with ASTM F 402-1988
2. Water service pipe must meet the appropriate NSF standard for potable water.
3. Type K or L copper may be installed underground.
4. Dimensions ratio (DR) 17 or less
5. Dimensions ratio (DR) 13.5 or less 6.

ASME B. 120.1-1983

As you may have guessed by now, water supply and distribution materials are available in plastic, steel, and copper pipe and fittings. Municipal water mains are typically constructed with cast iron. On the other hand, more commonly known as ductile iron.

Ductile iron is extremely heavy and is expensive to install and repair. When deciding what type of materials to choose for the installation of water supply and distribution, there are a few things to take into consideration.

First the cost of the material, the corrosiveness of the ground that it is going to be buried in. It is also important to take into consideration what the local codes allow. In addition, other considerations may be operating pressures, and conditions in which the pipe and fittings may be under. Such as environment, ground stress, and ease of installation.

In many cases, the life expectancy of galvanized and copper tube depends on soil and water conditions. Galvanized pipe is electroplated with zinc to produce a protective coating that resist corrosion or oxidation. Another consideration in selecting materials for use in the water supply and distribution is the possibility of galvanic corrosion, which can develop when dissimilar metals such as copper and galvanized are joined, or touching. This is also commonly called electrolysis; one way to prevent this is with the use of an insulating fitting such as a dielectric union, or brass coupling.

Electrolysis is caused by a weak electrical current that flows between the different metals. While on the subject of electrical current, I think this is as good as any time to talk about safety hazards when removing a ground strap. A ground strap is the section of wire that is run across the water meter. This sometimes requires the removal of the strap or wire in order to complete a re-pipe, if there is going to be a new connection at the house side of the meter.

There are two sides to a water meter; we have the "city side" or the incoming supply from the street, and the house side, which is just that. The ground strap or "jumper wire" will run from the city side of the meter to the house side. This connection is often done via a clamp in which the wire is locked down to. Typically, this is a single wire and can easily be removed with a flat screwdriver.

Now for the safety precaution, I for one have removed somewhere in the neighborhood of 1,000 without incident. Nevertheless, there is always the 1 in 1,000 that just may cause an issue. Remember as I said earlier, just when you think you have learned everything there is to know about the trade something new will come up. That will often leave you in surprising situation.

First, the cautionary and simple steps to avoid what could be a very costly expense. Anytime the removal of the ground strap is removed, you should always shut the main power off, if feasible. If the power at the fuse box cannot be shut off, the next best thing is to take a set of jumper cables and temporarily connect one end to the city side of the meter, and the other end to a point on the house side.

Once the ground strap is removed there is a condition known as a "floating neutral" this can occur when the earth ground is removed. In short, this is the equivalent of a racecar with a space shuttle engine attached to it. It is wildly unstable, but one thing is all most certain, there will be a crash.

On an older home that strap across the meter maybe the only earth ground of any kind. As with all trades, codes have been established for a reason, and usually we just need to look back in history to find out why. Electrical codes have changed throughout history as well as plumbing codes. It is common practice now a day to install not one but two earth grounds, usually on the outside of the home near the electrical meter. In most cases, the earth ground is an eight-foot long piece of solid copper, with a diameter of 3/8 of an inch. In newer homes, there are about two of the rods that are in the earth with six to eight inches exposed for clamping or grounding purposes. If the ground strap is disconnected without a temporary in place, the system could send a current throughout the electrical system and cause an electrical fire.
I have witnessed such an incident first hand. I would have never thought something like that could happen, but as I pointed out just when you have seen it all.

A plumbing contractor did exactly that. He disconnected the ground, inadvertently causing an electrical fire in an older home. One in a million chance, is what the electrical forensic said. By removing the ground strap, the plumber had overloaded the system and three wall outlets had caught fire. The only solution was to shut the power off at the electrical panel and call 911.

With some training on what precautionary steps he should have taken, this would not have happened. After seeing this first hand, I have included this as part of my four-hour continuing education class to help other plumbers understand that, this is a situation that could result in a fire. The water supplied by the city comes in from the supplied water mains, which are usually constructed of ductal iron pipe. From there, the main is tapped with a valve and copper or other approved water pipe is used to run onto the property. There is a valve installed once inside the property called a curb stop or "B-box" (buffalo box) at that point the water service becomes property of the homeowner. The curb stop is owned by the city, in most cities towns or villages.

The water distribution piping is of the same materials approved as the water supply. As for assembly of any pipe or fitting, the first rule is always cut straight and square. The next rule is the pipe end should be free of burs, clean and likewise for the fittings to be joined.

Basic Fitting Chart and Use

Fitting	Description / Use
90 degree ell & 45 degree ell	Used to change direction of the pipeline
90 degree street ell	An elbow with a male end and female end used to change direction of the pipeline 90
45 degree street ell	An elbow with a male end and female end used to change direction of the pipeline 45
Tee	A fitting with three openings used to make branches at 90 degree angles to the main
Coupling	A fitting used to connect lengths of pipe in straight runs
Cap	A fitting used to close the end of a pipe
Union	A fitting used to join two pipes or equipment the can be
Reducer coupling	A fitting used to join different pipe sizes

5 JOINTS AND CONNECTIONS

There are several joining methods used in plumbing, each section of pipe and fittings that are used every day by plumbers could fill another entire book. It is for that reason I have limited this book to the basics with the apprentice in mind. This is also a good time to point out this is the "C" section of the codebook.

The first thing to understand when we are joining pipe and fittings in a plumbing system regardless if it is sanitary or water distribution. It must be installed in a way as to make all joints gas, and watertight. In other words, NO LEAKS!

The first joint we will look at is the caulked joint, caulked joints should be used only in the drain waste and vent system. This joint is for use with cast iron pipe and fittings; the joint is firmly packed with either white or brown oakum. It is then filled with molten lead at least one inch in depth, and firmly caulked as not to extend more then 1/8 of an inch below the rim of the hub. Cast iron dates back to the 17th century and were used to distribute water throughout the gardens of the Chateau de Versailles. Cast iron proved to be a beneficial material for the manufacture of water pipes and was used as a replacement for the original elm pipelines laid in the ground earlier. Although there is recorded uses of cast iron in 1562 in Langensalza, Germany. It was used there to supply water to a fountain. The first full scale cast iron pipe system for the distribution of water was installed in 1664, the 15 mile long main served the Chateau de Versailles.

In 1845, the first pipe was cast vertically in a pit. By the end of the century, all pipe was manufactured by this method. Prior to this practice, it was typically cast in a horizontal mold. However, this resulted in an uneven distribution of metal around the pipe circumference. Typically, slag would collect at the crown of the pipe creating a weak section. Cast iron pipe was first used in the United States around the beginning of the nineteenth century. It was imported from England and Scotland to be installed in the water supply and gas lighting systems of the larger cities. One of the first cast iron installations was at Bethlehem, Pennsylvania, where it was used to replace deteriorated wooden water mains. The iron industry in the colonial United States was limited to the production of raw materials. This iron was shipped to England where it was re-melted and used to manufacture goods in the United States for the Revolutionary War.
Interesting enough, eight-foundry owner's signatures appear on the Declaration of Independence.

The decade of the 1890's marked the emergence of cast iron soil pipe manufacture as a distinct industrial activity. Cities continued to install water works and sewage systems at a rapid pace, and the total number of cast iron pipe foundries in the United States increased to 64 in 1894 and 71 in 1898.

The total in 1894 was divided equally between pressure pipe and soil pipe foundries. By 1898, there were 37 foundries devoted to soil pipe production. They were located in 13 states and had an annual melting capacity of approximately 560,000 net tons.

The production of cast iron soil pipe and fittings in the United States, which reached a peak level of 280,000 net tons in 1916. It slackened during World War I and totaled only 111,000 net tons in 1918.

Following the war, building projects, which had been deferred, were undertaken, and as construction activity increased so did the demand for building materials, including soil pipe.

During the early 1920's, the industry invested heavily in new plants and equipment. In Alabama, at Anniston, 5 new foundries were constructed which raised the city's annual output to 140,000 net tons and made it the largest production center for cast iron soil pipe in the world. By 1922, the nation's production of cast iron soil pipe and fittings had reached 357,000 net tons, and approximately 180,000 net tons or 50 percent of this total was produced in Alabama.
The spigot end is fitted into the hub end of the pipe; the joint must be seated squarely in order to achieve an even pour of molten lead.
This can also be seen in figure A.
Other methods of joining cast iron are no-hub connections and gasket connections.

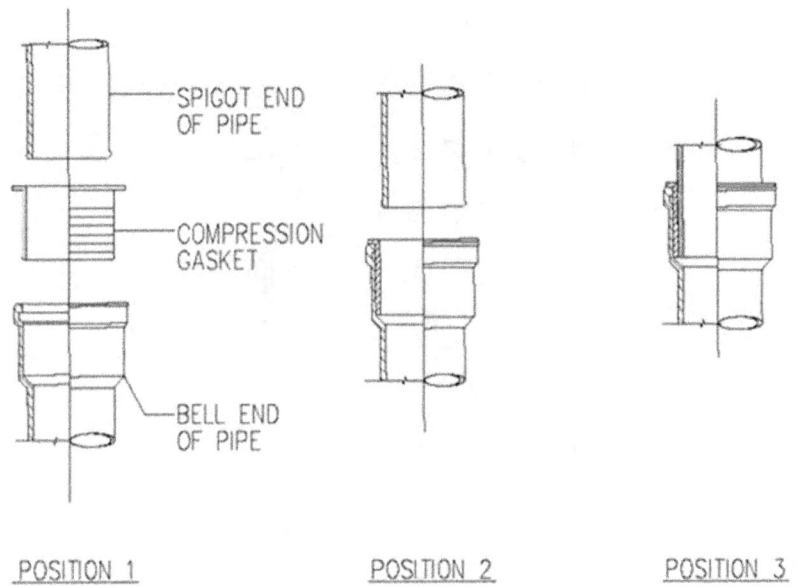

Position of application for compression joint

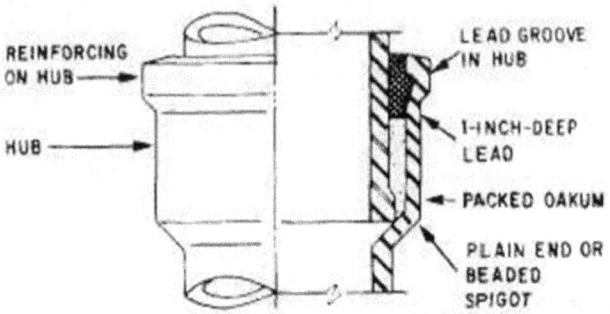

A. LEAD AND OAKUM JOINT

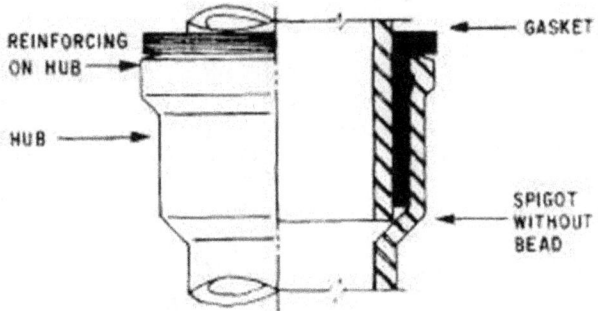

B. COMPRESSION JOINT

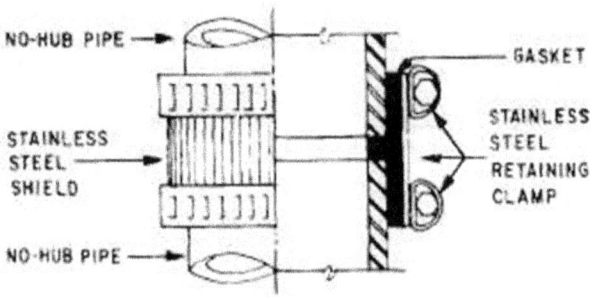

C. NO-HUB JOINT

Figure A. is an example of a lead – oakum joint in which molten lead is poured over an oakum packed hub to create a gas and water tight seal. Figure B. is an example of a slip seal or gasket connection, this is also known as a compression joint. This joint is most common in the use of the installation of a building drain underground. It is made with the use of a neoprene insertable gasket. Figure C. is an example of a no-hub connection. This example is connected with a shielded band and an elastomeric gasket. It is then tightened to torque specifications of 60-foot pounds. The examples provided in the diagram are the most common type of joints made when working with cast iron pipe and fittings. Cast iron is available in two weight classifications, extra heavy, and service weight.

The next method we will look at is threaded or screw pipe joints. All threaded joints must conform to American National Taper Pipe Thread (NTP). It is important to clean the burrs from the ends of the pipe before joining. In addition, wipe out any excessive oil after the threading process is complete.

Depending upon what type of application the threaded pipe will be used for, the excessive oil could be a problem once the system is active. NPT is the United States standard for tapered threads, in contrast to straight threads as used on a bolt, a taper thread will pull tight and, therefore, making a fluid-tight seal.

Pipe threads are typically a tapered self-locking thread produced on the outside diameter of a piece of pipe, tubing, or fitting. They are in the inside of the mating coupling, port, or fitting. As pipe is identified by the inside diameter of the pipe. Therefore, a ¼-18 NPT fitting has a
¼" diameter through hole. Pipe threads are designed to be assembled by hand (4 to 7 turns), depending on size, and drawn together an additional 2 turns by means of a pipe wrench or other mechanical device, locking the threads together. Prior to the 1800's, pipes and couplings were manufactured as matched sets with little regard to formal thread profile or size. In 1820, Robert Briggs of the Pascal Iron Works of the Morris Tasker Co. began work to develop the first "pipe threads".

In 1834 he made the first gauge to inspect internal pipe threads, a L1 threaded plug. In 1862, Mr. Briggs developed a mating threaded ring gauge for the external threads and published a standard for what was called the Briggs Standard Pipe Thread. By 1886, a large majority of manufactures threaded pipe to the Briggs standard, and acting jointly with A.S.M.E., they adopted it as a national standard.
Around 1905, The American Society of Mechanical Engineers, along with various government and military agencies, started the American Standards Association (ASA) with the purpose of developing the standards to be used nationally.

In 1919, the ASA, using the Briggs Standard Pipe Thread as it basis, created the National Pipe Taper (NPT) pipe threads. The organization also published the B2.1 standard, complete with all taper pipe and straight pipe specifications and gaging.

In 1927, to serve the automotive industry and create a self-sealing (Dry seal) thread form, the ASA B2.2 standard was created using a modified form of the NPT pipe thread called the National Pipe Taper Fuel (NPTF) pipe thread.

In 1961, the military, wanting a higher quality NPT thread created MiIL-P-7105, creating the "Aeronautical National Pipe Taper" (ANPT) Threads. The threads were the same as the NPT threads up to 2 ½" diameter, but required additional gaging and gage variation must be considered when gaging the threads. Copper is widely used for plumbing pipes because of its excellent corrosion resistance and safety. It is also very easy to work with, as it is malleable and easily joined by fittings or soldering. Copper plumbing pipe is available in four different types. Type K has the thickest walls; type L has thinner walls than K but thicker than M: type M has thin walls most common for use in water supply because of its low cost. Type DWV very thin walls and can only be used in the drain waste and vent system. This material is not designed for use under pressure. The surface to be soldered should be cleaned with sand cloth, as well as the fittings to be joined. Both the end of the pipe and the fitting must be fluxed. Use a flux that will dissolve and remove traces of oxide from the cleaned surfaces to be joined, protect the cleaned surfaces from re- oxidation during heating, and promote the wetting of the surfaces by the solder metal. As recommended in general requirements of ASTM B 813. Apply a thin even coating of flux with a brush to both tube and fitting. Another joining technology that has been used effectively for many years involves a hand tool designed to quickly pull tee connections and outlets from the run of the tube, this will reduce the number of fittings and soldered or brazed joints. It allows branches to be formed faster and usually results in a lower installed cost. This method may be used for plumbing heating and refrigeration.

Portable hand tool kits and power operated equipment are available that produce lap joints for brazing. The system can be used with Types L, K, or M copper tube to from ½ inch to 4-inch outlets. Another form of joining copper that has grown in popularity over the last several years is the press method. Press-connect joining of copper and copper alloy tube is very fast, economical, and most importantly, does not require any heat source.

The press-connect method, sometimes called press-fit, was patented in Europe in the late fifties, and continues to be used successfully there today. The method and associated fittings and tools were introduced in the United States in the late 1990's. Since then, there has been growing acceptance, and those using the method have experienced excellent results.

Press-connect joining takes advantage of copper's excellent malleability and its proven increased strength when cold worked. The joints rely on the sealing capability of a special fitting that contains an elastomeric gasket or seal, and the proper use of an approved pressing tool and jaws. Typical ranges of pressure-temperature ratings for these no-flame joints can be found in the chart below.

Joint type	Pressure / Temperature Range
Press-connect ½ - 4 inch pipe	200 psi / 250 degrees
Press-connect high psi ½ - 1-1/8	700 psi / 300 degrees
Push-connect ½ - 2 inch	200 psi / 250 degrees

Push-connect (shark bite) is another popular method of joining copper, like the press method the push-connect will join copper and alloy copper tube is also fast, economical, and requires no heat or flame.
However, unlike most other joining methods, no additional tools, special fuel gases or electrical power are required for installation.

Push-connect joining utilizes an integral elastomeric gasket or seal and stainless steel grab ring to produce a strong, leak free joint. There are two common types of push-connect fittings. Both create strong, permanent joints. However, one allows for easy removal after installation to allow for equipment service, while the other cannot be easily removed once the fitting is installed.

The next joint we will look into is the flare joint, while copper tube is usually joined by solder, brazing, or one of many other methods, here are times when a mechanical joint is required. Flared fittings are an alternative when the use of an open flame is impractical. Water service applications generally use a flare to iron pipe connection when connecting copper tube to the main and or the meter. In addition, copper tube used for fuel gas, LP, propane gas or natural gas may be joined utilizing flared brass fittings of single 45-degree flare type.

A flare joint should be made with the appropriate tool such as those supplied by a number of tubing / piping toll manufactures.

Make sure to use the tool that has the appropriate flare angle, commonly 45 degree. The tool usually consist of flaring bars with openings for various tube sizes and a yoke that contains the flaring cone and a clamp to grip the flaring bars. When flaring Types L or K copper tube, annealed or soft temper tube should be used. It is possible to flare Types K, L or M rigid or hard temper tube, though prior to flaring it is usually necessary to anneal the end of the tube to be flared. The copper tube must be cut square using a tubing cutter. After cutting, the tube must be reamed to the full inside diameter leaving no inside burr, tube that is out of round prior to flaring should be resized back to round. Below *(fig. f)* is an example of a flare joint. Correct method *(Fig. c)* of a flare joint.

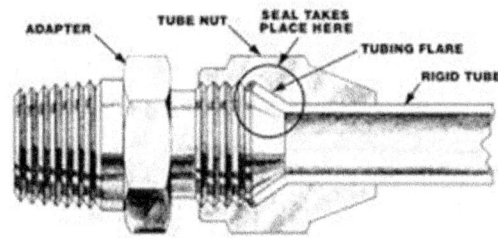

Figure F.

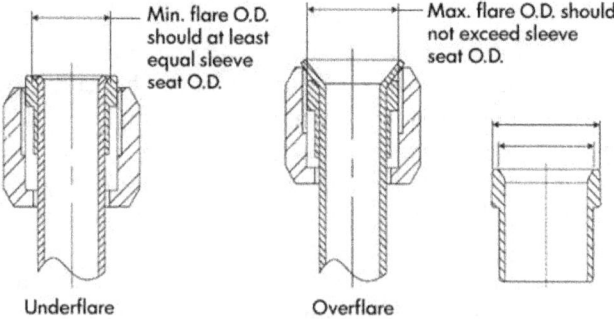

Figure C.

The compression joint or fittings have two or more ports or entry points for the connection of pipes or fixtures. They may be compression only, compression or screwed only. The compression port is used for connecting a pipe and the screwed port (if included), known as a union, connects to a fixture or device such as a radiator, furnace, spigot, or pressure gauge.

The screwed section of the fitting may have male or female threads. Fittings with compression ports only, are used to join two or more pipes together, and are simple to use. The ferrule as shown in the drawing below *(fig1.)* is the part of the connection that creates the seal. By tighten down on the threaded nut the ferrule will deform and compress around the tubing.

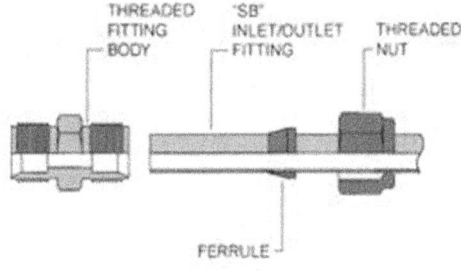

Figure 1.

Compression fittings can be used to connect plastic or copper and can be found on most plumbing fixtures, such as the angle or straight stop at most faucets, and water closets. You might even find a compression connection on most residential icemakers. While on the subject of icemaker kits, the saddle valve that is sold in the kit; is not allowed under the Illinois Plumbing Code. Repeatedly an apprentice will say, "If that part is not allowed to be used how is it that, the big box store is can sell it." The answer is simple it is called free trade. Just because a part is not allowed to be used under, the plumbing code does not give the state the right to "ban" a part from being sold. It is up to us as plumbers to know the difference and install the correct part as outlined in the plumbing code.

There are many types of connections and methods to joining pipe, the one that is often mistaken and installed incorrectly are the no-hub, fernco, or mission couplings. These have a particular use and more often than not used wrong. When joining plastic to a metallic material, a no-hub adapter must be used to ensure the fit and seal of the two materials. In the 1960's no- hub, couplings were the next best plumbing innovation to hit the market and transformed cast iron drainage pipe installations. Although no-hub couplings were designed for cast iron, they are also a handy way to join or repair ABS, and PVC. They have also become the standard on most renovation projects, when joining a cast iron pipe with PVC. No-hub adapter in *fig. 15* is an example of the proper adapter to be used when joining PVC to metallic material.

PART NO. 119

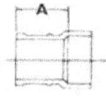

NO-HUB ADAPTER
(Adapts C.I. No-Hub spigot to DWV spigot)
SPIGOT X HUB

SIZE	A	B	C	D
1½	1⅝			
2	1⅝			
2 x 1½	1⅝			
3	1¹³⁄₁₆			
4	1⅞			
4 x 3	2¹⁄₁₆			

PART NO. 122

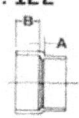

SPIGOT ADAPTER, CAST IRON
(Adapts cast iron spigot to DWV spigot)
HUB X HUB

SIZE	A	B	C	D
2 x 2	⅜	2⅜		
3 x 3	⁷⁄₁₆	2⅝		
4 x 4	½	2⅞		

AVAILABLE IN PVC ONLY

15

Another method or joint that is commonly used in plumbing, especially for the installation and or removal of equipment is the union. This joining method allows pipe of dissimilar metals to be join by means of a dielectric union. Unions may be used in the drainage system and venting system when accessible located above ground. This pertains primarily to the trap seal and the inlet, outlet of the trap. Unions must also, as found in Illinois Plumbing Code (890.350), be within five feet of regulating equipment such as water heaters, pumps, water conditioning equipment and conditioning tanks, and similar equipment which may require service or replacement. There are, of course, prohibited joints and connections when applying joint in the drainage system. A connection within the drainage system cannot have a raised edge, shoulder, or reduction in size within the direction of flow. No connection or fitting that obstructs flow may be used in the drainage system.

The T-DRILL System of mechanically formed tee connection was invented in 1967 in Finland by a plumber/engineer named Leo Larikka.

Mechanical formation of tee connections had been around for many years, but Larikka developed a more sophisticated method that became the most widely adapted system of its kind in the world. In fact, in Scandinavia they do not put in tees- they put in "Larikkas," just as we use "Kleenex" tissues and "Xerox" documents. Patents were not pursued until the early 1970s, though the second- and third-generation T-DRILLS have since been patented.

As plumbing wholesalers are rare in Finland, Larikka developed his invention out of necessity. Lacking the right size tee fitting, he came up with a way to create what he needed out of the tube itself. Before this method of installation could be widely used in North America, national, state and local codes had to be rewritten after suitable testing. Now, 15 years after introduction here, most plumbing codes have complied with and the method complies with ANSI B 31.5 (ASME code for pressure piping) and being used by over 5,000 plumbing, mechanical and sprinkler firms.

6 TRAPS AND CLEANOUTS

What is a Trap?

Before we discover the many traps that are in plumbing, let us discuss the basic principles of what a trap is and what job a trap is responsible for. First off, every plumbing fixture must have a trap; some are visible while other are hidden. The most obvious trap is located under your sink- either a kitchen, lavatory, bar sink or any other sink you can think of.

In order to understand the importance and significance of the p-trap we will need to go back in time once again and see just how this important plumbing design is. Alexander Cummings, in 1775 had invented a new device called the "S" bend and was later called the "U" bend.

The trap or fixture trap is in plain sight, it is the "j-bend" portion of the waste just below the strainer. They are commonly called a p-trap.

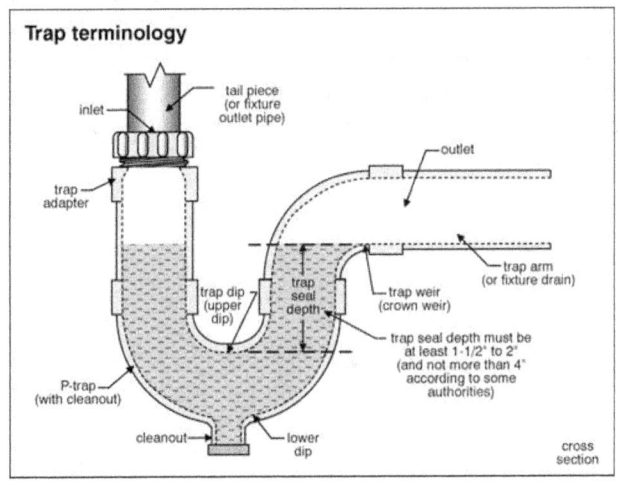

Let us take a closer look at the p-trap as in fig. 4 and first understand the proper terminology of the part that make up a p-trap. The tailpiece is the short section of piping that is attached to the strainer or pop-up assembly directly above the p-trap. This also the trap inlet from which water enters the trap by gravity. Water enters into what is called the trap seal area, this portion of the trap is designed to hold or retain a small portion of liquid in it, thereby creating a seal. This is of great importance because without a water seal, sewer gases can enter into the living space of a building.

As water is entering the trap and being used by flowing out through the outlet portion of the trap arm, the seal is replaced by water flowing into the trap. This also known as hydraulics or fluid power/mechanics. Hydraulics in plumbing is a part of physics called fluid mechanics, plumbing which deals with the flow of fluids in pipes, is a practical application of hydraulics. Hydraulic principals are based on the chemical, physical, and mechanical properties of water. Some properties that include density, viscosity, water temperature, and pressure. Plumbing deals with two types of fluids, gas, and liquid. The main difference between these two fluids is that gas is compressible. Liquids, mainly water in plumbing applications, are essentially non-compressible at room temperature. Water flows in pipes by gravity from a higher to lower elevation. The difference in elevation, or water level, in a system is called "static head." The same is true for an inactive p-trap; once again, the environment plays a part in the rate in which a trap will evaporate the water from the trap seal. This will occur most often in a seldom-used floor drain.

Evaporation can occur because the water is exposed to air causing the trap to "dry-out." One way to protect the water seal is to add a small amount of vegetable oil. Another common way the floor drains water seal is protected; is by using a device called a "trap primer." This can been seen in Fig.6 Over time, in infrequent used floor, drains can dry out, leading to the release of sewer gasses into the environment. The trap primer mitigates this problem by injecting water, either directly or indirectly, into the trap to maintain the water seal. There are several trap primers available to the plumber, and just as many manufactures. One in particular is an automatic primer that has a physical connection to the potable water supply. This type of device, if not fitted with a backflow preventer, must have one installed at the point of connection to the potable water source.

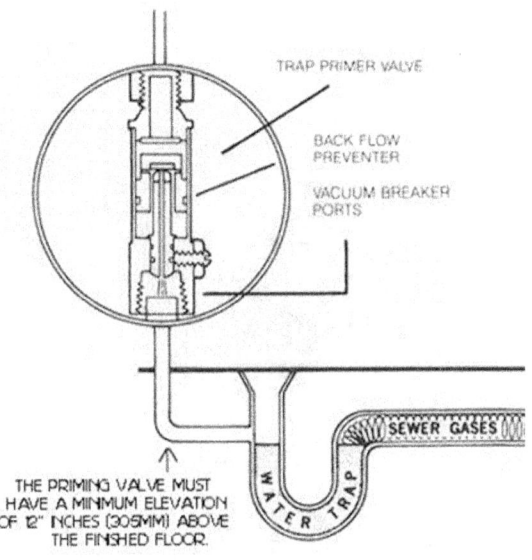

Figure 6

Trap primer devices are often installed during new construction or a remodeling project. Most common in a commercial application, these devices are used in conjunction with a specialty fitted floor drain in which the floor drain has a tapped or threaded connection just below the floor drain grate or finish. The water piping is then run underground from the priming device to the floor drain. The automatic devices must been installed in an accessible location, to allow for maintenance purposes. These devices will also need to be of an approved agency, which will bear the agencies stamp such as ASME, NSF, or ANSI and must be of lead free material.

Trap Types

Let us look at p-traps that are approved as well as traps not allowed to be used in Illinois. First off the traps that are approved for use are:

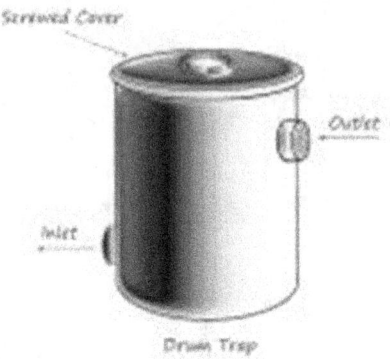

Drum Trap

Owners of an older home may be all too familiar with this style of trap. It holds true for any plumber who has ever tried to clear or "rod" through one of these. They can be a challenge especially for those just starting out in the plumbing trade. The truth is this is a "legal" trap in Illinois.

This style trap is often found alongside of a bathtub, with the clean out portion or cover of the trap located flush with the floor (*Fig. 3*) the challenging part of this trap is it is not always visible. They are sometimes hidden within the floor, or have been covered over by floor tile.

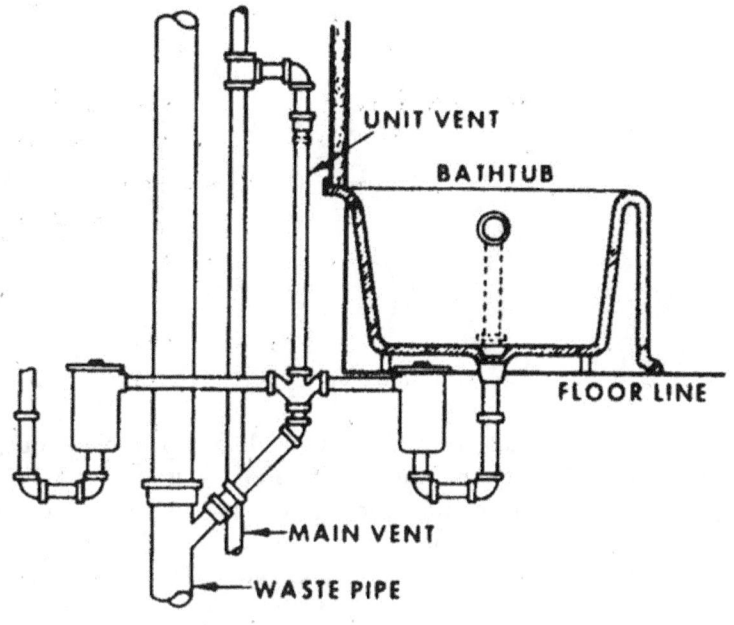

Figure 3

In the case that the cover is exposed, it has usually corroded and cannot be removed. The other is that this is often connected to galvanized or screw piping. The weakest part of any threaded pipe is located at the threads. If you attempt to "muscle" the cover open, you run the risk of turning the entire trap, thereby breaking the threaded connections. For the apprentice plumber who may have this in a customer's home will most likely attempt to clear the bathtub drain by gaining access through the overflow of the tub. This method is fine right up until the cable enters the drum trap. It is physically impossible to manipulate the cable into the downstream side of the drum trap. The next and obvious solution is to replace the drum trap with a new p-trap. Of course it is also a good idea to replace the waste and overflow of the bathtub as well, after all chances are that section of piping is as old as the drum trap.

Drum traps are available in many different types and configurations. Some may have the inlet located on the very bottom of the trap while the outlet is located at or near the top. No matter the pipe configuration, they all operate relatively the same. They were originally designed to act as a hair trap, which any plumber can tell you that part of the design works fairly well.

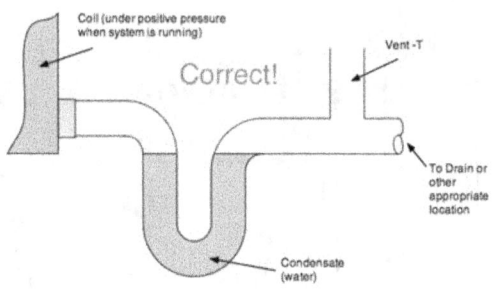

Running Trap

The running trap, which is an approved trap in the Illinois plumbing Code, is required to have a vent. This trap is much more prevalent in the HVAC world as it is often used in the condensation discharge of a humidifier. The running trap; also called an in-line trap, or house trap, is also utilized during the installation of trench drains in a commercial application. The vent must always be installed downstream of the trap, and must have an accessible clean out both up and downstream of the trap. The running trap can help to meet one code challenge that has come up in a few situations where a standard p-trap could not be used. The code that I am referring to states that traps must be protected from freezing. The situation that this came about was in a bath remodel job. The bathtub waste had sat just above a cantilever; which in simple terms means the waste including the trap had sat outside in an unheated space.

Of course, the concern was, what is going to happen to the water seal in the trap once the outside air temperature drops below 32-degrees. The funny thing was the architect had drawn the plans showing the trap outside in the unheated space, the plan examiner at the village approved the drawing, and the general contractor had constructed the cantilever; all had no concern as to the p-trap and possible freezing that could occur. As the plumber doing the installation, I had brought up to the general contractor, and the plumbing inspector, "hey guys I think we have a problem." I had a concern because I guaranteed all of my work, and was not about to install something knowing full well it would fail in the next few months.

I did what any other plumbing contractor would do, I asked the plumbing inspector to site visit the job, and get his opinion on the best way to protect the p-trap. Together we came up with the solution of running the waste from the tub into the heated area of the building and installing a vented running trap. The running trap had come through and saved a lot of money, by keeping the design of the building and not needing additional heating ducts.

To sum up, the traps that are approved by the department for use in a plumbing system besides the standard p-trap, the drum trap in example one, and the vented running trap (example two). Now let us jump in and look at the prohibited traps that cannot be used.

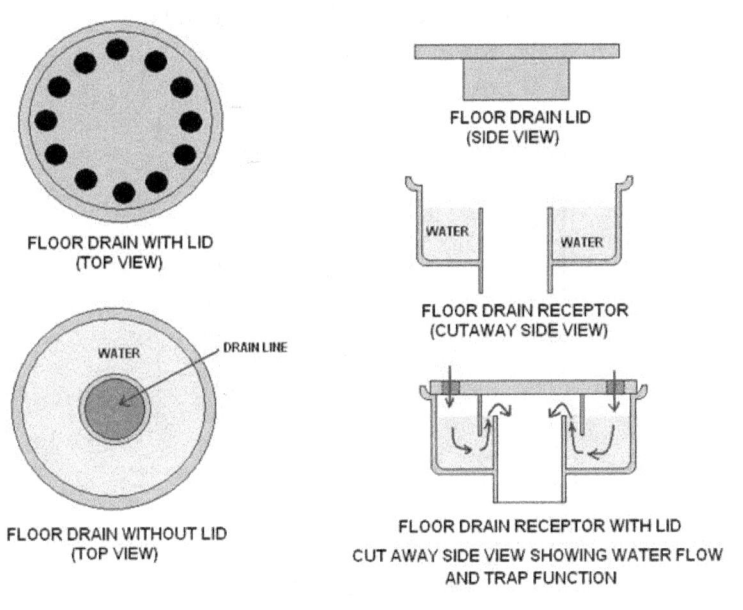

Bell Trap

The bell trap can be found in homes dating back to the early 1900's these trap would have been used as a basement floor drain, they are prone to clogging easily and are far less sanitary than a traditional p- trap.

S-Trap

This style trap once again is typical of the traps used in the early 1900's. These are most commonly found under kitchen sinks and lavatory sinks. They are often connected directly to the drainage system without the use of a vent. The lack of venting may result in the loss of the trap seal and the fixture may not drain properly without adequate airflow.

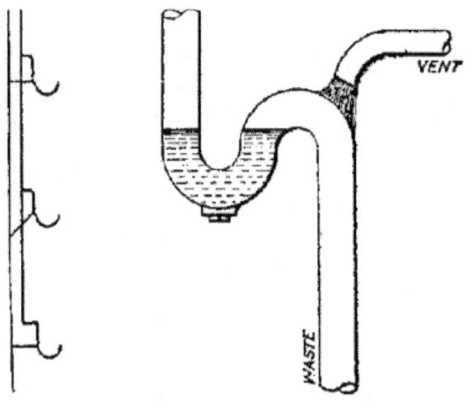

Crown Vented Trap

This means any trap that is vented less than two pipe diameters in distance from the trap is considered crown vented and is not allowable. The distance is measured from the trap weir, which is the trap elbow, to the vent connection. It is also noteworthy to bring up just a few additional comments regarding p-traps. Such as when the subject of fixture trap is discussed, it is important to remember all plumbing fixtures are required to have a trap. This should not be confused with fixture's having an integral trap as with water closet's, and urinals. In addition, no fixture shall be double trapped. Traps shall also have no moving parts, non-adjustable trap seal and must be of smooth and uniform interior. The trap seal must be a minimum of two inches and not more than eight inches. In Illinois, a building trap, or a trap located in the building sewer is not allowed.

Pipe Cleanouts

Rod station and sewer line access are just two of the many terms used to describe what is commonly known as a cleanout. A cleanouts main purpose to provide an accessible means to maintain a sewer system. Cleanouts provide access to clogged pipes, on a municipal level they may also be a manhole in the street.

With that, let us discuss the appropriate distances for clean outs, and at what point a manhole is required. First, in a four-inch sewer pipe, the maximum distance a pipe run can be without a full size cleanout is 50 feet in a horizontal direction. This is also the case with a pipe size of 1- 1/2, 2", or 3"

Now here is when the codebook may get a little confusing; it also tells us that in a sewer pipe of over 4-inches up to a 10-inch pipe you are allowed to run in a horizontal direction of not more than 100 feet.
Really, all this means is if you run a 4-inch sewer pipe at a distance of 51 feet you are required to install at least one cleanout. That cleanout by the way must be the same size as the sewer pipe; in other words, you cannot install a three-inch cleanout on a four-inch sewer pipe. If you are installing piping of ten-inch in diameter or larger you may run in a horizontal direction of not more than 150 feet. With the installation of large diameter, piping a cleanout is of no use, and therefore will require the installation of a manhole. So where is a cleanout required in relationship to an interior plumbing system? First off at the base of every stack, in addition no higher than 48 inches from the finished floor, also at every change of direction. Let us get a clear picture of what would constitute a change in direction.

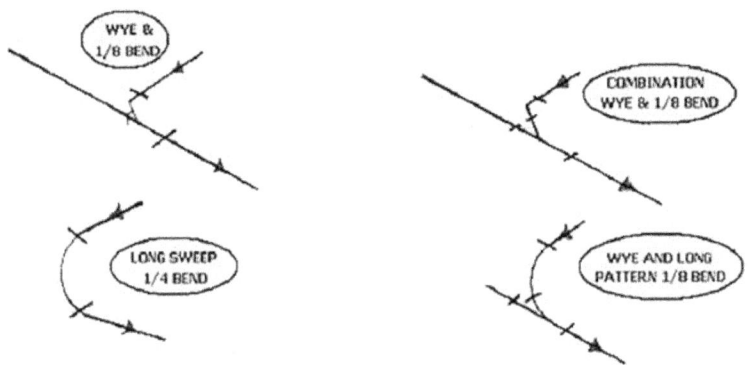

Figure 8

This direction change could also be from vertical to horizontal. However, the cleanout is most commonly installed at the end of such run. In simplified terms, if the change of direction is, greater than 45 degrees from the intended path a cleanout should be installed.

One misconception about building codes and cleanouts is that a cleanout is required five feet outside a buildings foundation. The truth of the matter is; is that the codebook tells us the cleanout can be inside or outside of the building. In addition, the cleanout, when installed outside of a building, should be installed with fittings that will allow a rodding machine to penetrate both upstream and downstream of the sewer pipe.

Fixture trap, readily removable and without disturbing concealed plumbing or requiring the fixture to be removed, is acceptable as a cleanout equivalent, if there is no more than one 90 degree bend in the line being rodded. A water closet is not a cleanout equivalent. The reason that I bring this up is I have had that very question asked, "Is a floor drain a cleanout equivalent?"
The answer is no, because technically a floor drain has two 90 degree bends in it.

7 INTERCEPTORS AND BACKWATER VALVES

Types of Interceptors

Interceptors and backwater valves have many different applications and as many installation guidelines. The most common applications for grease interceptors and traps can be found in all restaurants in Illinois. We will begin this chapter with an overview of the basics, and I will explain exactly how to size a grease trap for your project. Let us begin with the basics. Our codebook tells us kitchen equipment must be drained into the sanitary sewer after running through a grease trap. The basics behind the design is very simple. The idea is to capture the solids (grease and food waste) prior to entering into the sanitary system.

If the kitchen equipment, including prep sinks, vegetable sinks, and dishwashers, did not first discharge into a grease trap, the sanitary sewer system would literary be choked by the thick food waste and grease. Not to mention the public sewer system would virtually stop flowing. There are several manufactures of grease interceptors, as well as gas and oil separators, which we will look at as well later in this chapter. First, let us look at one manufacture as a reference, only because our codebook points out that the interceptor must be constructed of corrosion resistant materials, which means polyethylene.

Steel interceptors or not corrosion resistant and will at some point fail. The manufacture we will look at as a reference is Schier products. They offer several types of interceptors from small to large and chemical holding tanks for corrosive materials. Now not to sound like a commercial, but their traps are lightweight in addition, very durable. I have at one time or another installed this brand as well as others, made of steel and larger concrete basins, and It's my personal preference to go with the light weight one every time. Not only that; the interceptor will last a very long time, which will make your customer very happy as well.

Schier manufacturing offers several interceptors ranging from a flow rate of 15 up to a very large unit with a liquid capacity of 1,500. The larger model is being used more frequently now than ever before. This has a lot to do with the changing codes and the sanitary reclamation district having jurisdiction of an area.

Most restaurants are now being required to have all plumbing fixtures including floor drains, floor sinks, and equipment to have a separate "grease line" for the entire kitchen. In smaller buildouts, having only one triple compartment sink you still may be able to install a smaller trap. As always, it is a good idea to check with the local plumbing inspector and the sanitary reclamation district for clarification.

The Illinois plumbing Codebook also states that all waste lines and drains carrying grease, fats, or culinary oils shall be directed to one or more interceptors. All interceptors shall have watertight covers securely fastened in place. The grease trap that is either specified in the drawings or if you the plumber are going to take on that responsibility to determine what type of interceptor to use be sure of one thing first, it must be approved by the department and or an approved agency as listed in chapter four.

Plumbing and Drain Institute (PDI) should have their stamp of approval on the interceptor. Bear in mind that the plumbing inspector can reject the interceptor after its been installed based on that reason alone. The diagram on the next page shows the basic operation of a grease interceptor. Regardless of size, all grease traps work and operate on the same principal; food waste and grease materials will float to the top of trap while the liquid waste travels through the trap and out to the sanitary sewer system.

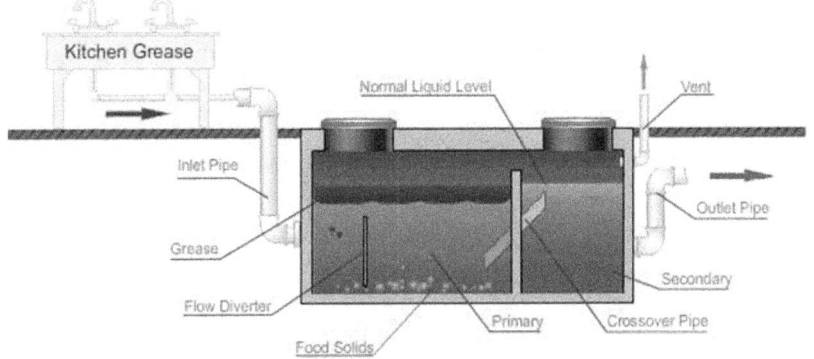

Now on the commercial vehicle side of the chart, buildings, which conduct vehicle maintenance, gas stations, with grease racks or pits, including oil change facilities, shall be provided with floor drains or trench drains. They must be connected to a gas and oil separator. If these buildings are connected to a private sewer system, the floor drain and or the trench drain can be connected to a holding tank in lieu of a gas and oil separator. Where trench drains are used to carry gas and oil waste to an interceptor, the trench drain must extend the entire length of the stall or work area. For multiple work stalls one trench drain is required for each stall.

A residential garage is exempt from this requirement as long as there are not more than five bays. These are often referred to as a triple pit and are in every car dealership across Illinois and most likely the country. They are in many ways similar to the grease trap in a restaurant. The gas and oil separator operation is the same as the grease trap. It is designed to capture the sludge from oil dry, oil, and other petroleum products that may end up on the floor during vehicle maintenance. As with vehicle maintenance the triple pit will also be installed in airplane hangars, bus garages, and other garages where vehicles may be stored or serviced.

Z886 Installation Specification

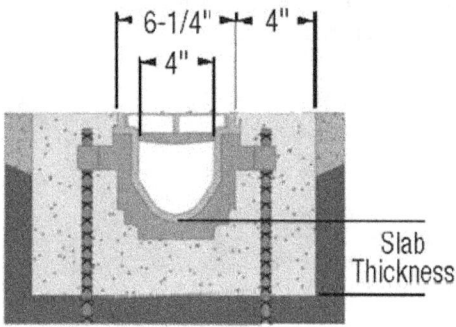

A typical side view of a trench drain used in a vehicle maintenance garage.

Trap Sizing

Below is a drawing of a typical garage triple basin or gas and oil separator. One main difference in the installation of the triple basin shown below when compared to the grease trap, is that the triple requires each basin to have its own vent as depicted in the drawing its shown as a two inch opening. When placing and ordering a basin as this, it is important to know exactly how it is going to be installed once at the job. The most important thing to know is what is the invert or the elevation of the inlet is and what is the position of the vent openings will be. North or south is the only options for this.

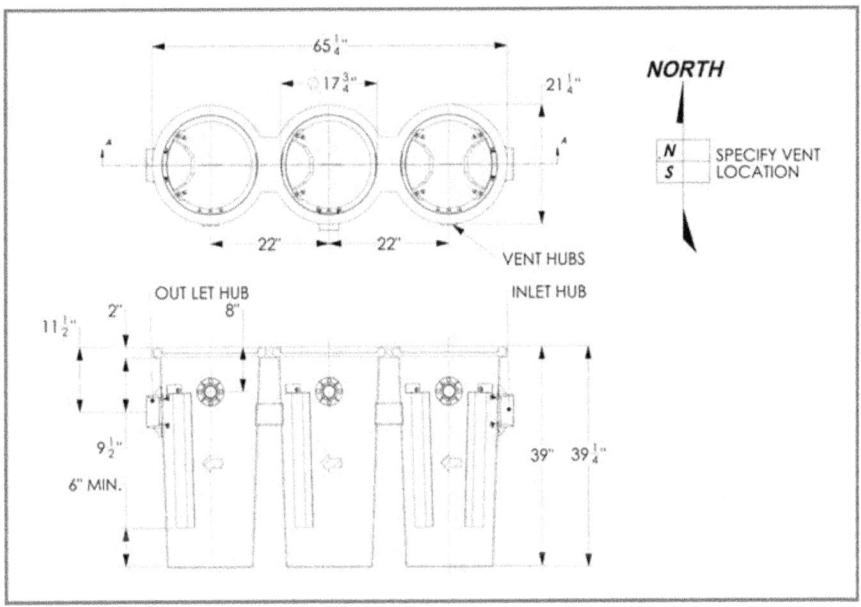

As for sizing this type of interceptor, it is important to remember the square footage of the area is used to calculate the size of the receptacle. Occasionally a pump or "lift station" may be required during the installation. It is critical to have all the necessary measurements when sizing this type of interceptor. If a pump or lift station is to be utilized during the install the lid and crock should be explosion proof.

As far as how to size a grease trap, I have kept the formula as simple and as easy to explain as possible. In order to calculate our grease trap we will use the example of the triple sink as listed below.

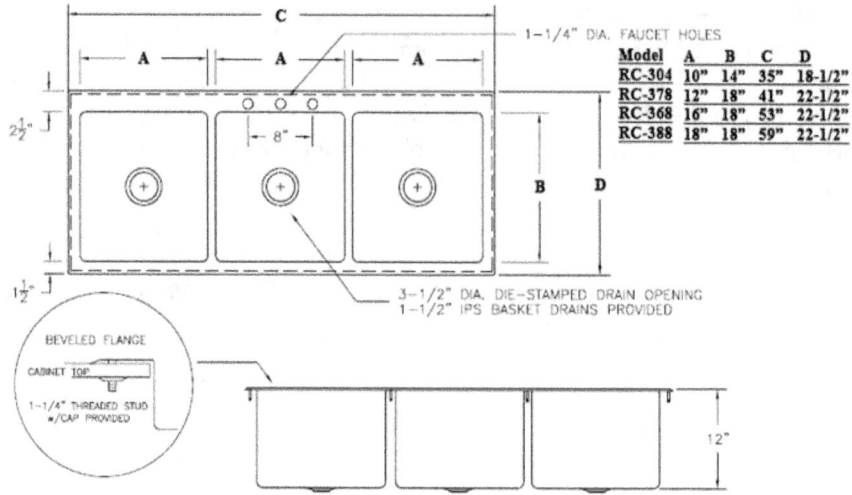

From the drawing we can see that we have dimensions listed this is our starting point to determine the size trap that will be required. First what we need is the length of the basin; we can tell from the chart on the right side that dimension (A) is 10 inches.

The next dimension we need is the width of the basin. From the chart, we can see the width (B) is 14 inches. The last dimension we need is the depth of the basin, which is listed above as 12 inches.

The formula that we need is (L) length x (W) width x (D) or written out as 10 x 14 x 12 = 1,680. What we have just figured out is the square cubic inches of one sink basin. However, we have three in total; next, we take the 1,680 and multiply by three, which gives us a total of 5,040 cubic inches.

Next what we need to do is convert the cubic inches into gallons of fluid or water. For this, we need to divide our 5,040 by 231. The reason we divide by 231 is that, this is the equivalent of how many cubic inches there are in one gallon. Now we have a number of 21.8181818. We need one more piece of the puzzle before we can call in our order to the supply house.

If we look at our codebook, we can see under 890.510 (a). 2. That our trap is required to hold 50 percent of the holding capacity. We will need to multiply once again by .050 to get our calculation of 10.9090909 or 11 gallons.

Based off the information provided from the triple sink specification, we now know we will need to order an 11-gallon capacity grease trap. I will list the formula once again below so that it will be slightly clearer.

(L) X (W) x (D) = 1,680 x (three sink bowls) = 5,040 divide by 231 = 21.8181818 x .050 (50% capacity) = 10.9090909 (11 gallons)

10" x 14" x 12" = 1,680 x 3 = 5,040 divide by 231 = 21.8181818 x .050 = 10.9090909 (or 11 gallons)

If the trap is to sit under the floor of the fixture, the holding capacity changes from 50 percent up to 60 percent. The rest of the formula is the same as described above.

When we need to calculate the size of a gas and oil separator the one and only thing we need to know is the square footage of the garage area. In addition, or more to the point of the square feet of where vehicles will be kept stored or serviced. Gas and oil separators are calculated based on square feet and not the amount of trench drains or floor drains that will be draining into them. Once the square footage is determined, you simply call the manufacturer of the basin. I have always used AK Industries for all my gas oil separators. Their team will ask a few basic questions and calculate the size of the basin. They design and build the basin based from the information you provide.

Backwater Valves and Flood Control

A backwater valve is sometimes confused with a backflow, which is incorrect. The backwater valve is device installed within the sanitary sewer line. A backflow is a device, installed within the domestic water supply piping. In fact, a backflow device is a complex safety valve with spring-loaded checks some have a relief port, some have multiple moving parts. A backwater valve has one main component, a flapper, which allows water to travel in the direction of flow. It will thereby close during a rain event.

Backwater valves are somewhat a popular device within the city of Chicago. This is because the city has what is known as a combination sewer system. What this means is the sanitary and the storm pipe system is one in the same. The device is installed within the sanitary piping usually at or near the interior of the foundation. It is recommended for those who live in a flood prone area, and is designed for use on a gravity type sewer system. There are two basic types of sanitary systems. Type one is known as an overhead that is a sanitary waste system that exits the building through the foundation.

Type two is a system known as a gravity sewer, which means the sanitary waste system exits the building underground, and under the foundation and footing. The backwater valve will always be installed on the type two system, which is the gravity sewer. It is important to know this device should be installed in an accessible area, as it will require some maintenance.

Usually the best practice for the installation of such a device is to install it within a pit, such as a poly sump pit or equal. With that, a secure cover should be placed over it as well, especially if the device is going to be installed within a living space such as a garden apartment or finished basement.

These devices are manufactured in either PVC or in cast iron, once again knowing the local code could save you time and money prior to installing a backwater valve. They are also available in a variety of sizes; remember if the sanitary sewer or building drain is 6 inch then the backwater valve must also be 6 inch. How does a backwater valve work? It can be installed as in the basement at the exit point of the sanitary waste piping. The device has arrows, which are in the direction of flow. If it is installed backwards, nothing in the building will drain out. The main component to the device is a flapper that is hinged usually on the topside of the unit.

This allows wastewater to flow out of the building. During a rain event where the sewers are overwhelmed with water, the flapper will close as water is forced in the opposite direction. It is a very simple design yet very affective when controlling water from entering back into the building.

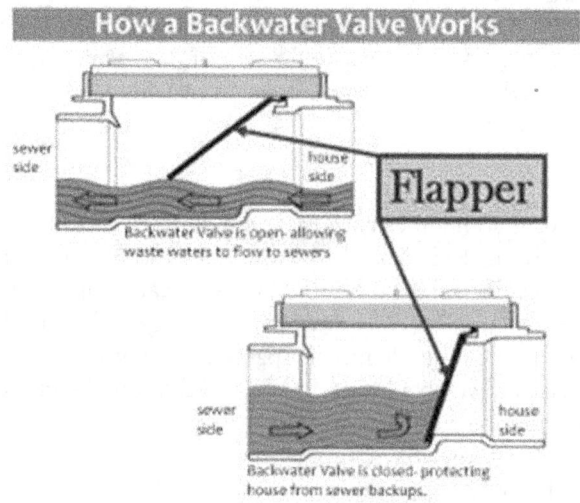

When these devices are installed, it is advisable to install a cleanout downstream of the unit as to allow access to the sewer for maintenance purposes.
Another alternative to the backwater valve is a device called a clean check, which can be installed outside of the foundation of a building. Unlike the backwater valve, the clean check can be readily accessible through a cleanout typically located in the yard. Once again, with the installation of a clean check it is recommended that a cleanout be installed downstream of the device.

Clean Check installed in a sewer

Now for the most expensive model of controlling floodwater. Most commonly referred as flood control system. As with the backwater valve and clean check, once the flapper is in the closed position, water will not flow back into the building and with that, water will not flow out either due to the forces on the downstream keeping the flapper closed tight.

A proper flood control will be installed inside of a manhole. It will include a backwater valve and an ejector pump in order to force wastewater from within the building downstream of the backwater valve. The manhole and equipment is installed outside of the buildings foundation.

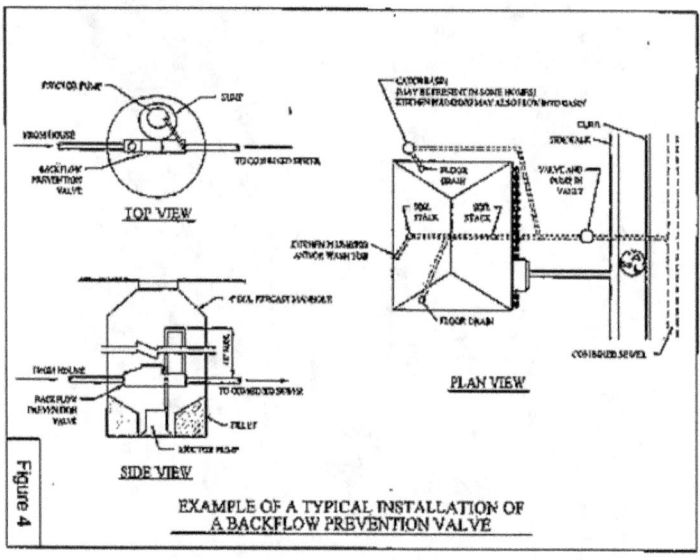

Figure 4 is an accurate representation of a flood control system in most installations. This can be completed within one day and has a heavy price tag of around 25,000 dollars to complete.

8 PLUMBING FIXTURES

The Basics

Everyone is familiar with basic plumbing fixtures, and how they operate. Plumbing fixtures include the basic water closet, lavatory, and bathtub. From the original water closet, which was at one, point nothing more than a hole in the ground. Humankind, with revenue as the main driving force, has developed new and far more technically advanced plumbing fixtures. Quality of plumbing fixtures shall also be of approved design and constructed of approved materials. Before the standardization of plumbing codes around 1928, there was very little worry about materials or agency approvals.
The international Association of Plumbing and Mechanical Officials developed the UPC or uniform Plumbing Code (IAPMO) to govern the installation and inspection of plumbing systems. At the focus was to promote the public's health, safety, and welfare as Thomas Crapper had done in the late 1800's. During the mid-1920's, there was rampant sanitation and disease related cases, which prompted a group of Los Angeles plumbing inspectors to create a minimum code of standards. With that, there was also a lot of disorder within the industry. The outcome was that in 1945 the Western Plumbing Officials Association adopted the Uniform Plumbing Code.

Later changing its name in 1966 to the (IAMPO) as we know it today. With the publication, reaching a milestone in 2003 with, for the first time in history, a plumbing code was developed through a true consensus process. Contributors to the content of the code were from every segment of the industry, including such diverse interest as consumers, enforcing authorities, installers, insurance, labor, manufacturing, and research/standards/testing laboratories, and experts. The codebook in Illinois is made up primarily of the Uniform Plumbing Code. With that, let us focus just a little on what and how this relates to plumbing fixtures. We know that all plumbing fixtures must have smooth, impervious surfaces and be free of defects and concealed fouling surfaces.
Used plumbing fixtures maybe re-used as long as they meet the minimum standards. If an overflow is to be a part of a plumbing fixture, and keep in mind an overflow is not a requirement.

It must be designed so that the standing water in the fixture cannot rise above the overflow when the stopper is closed. In other words, the overflow must be a part of the waste piping.

Installation methods, all plumbing fixtures shall be aligned, level, and secure. Corrosion resistant brass screws or equal must be the fasteners or means of securing a fixture to either a wall or floor. Support used must be durable and strong as to not transfer any strain to the fixture. When connecting a water supply to a plumbing fixture, only a potable water source may be used. Approved plumbing fixtures must be used in the manner in which the manufacture intended its use. When plumbing is installed in a public building, it must also meet the Americans with Disability Act (ADA) codes. Which is a whole other codebook that must be followed, when installing or designing a plumbing system for public use.
One thing that should be brought up is when it comes to the installation of plumbing fixtures it is important to remember that with each plumbing fixture; within the box will be a set of installation instructions. Those installation instructions, if the fixture has been approved for use will trump the plumbing code.
As long as the department has approved the fixture you are installing, so have the installation instructions as outlined by the manufacture. They should be followed closely and accurately.

The Men Responsible

One person that is synonymous in the plumbing world with his innovations and inventions is Al Moen. Al Moen holds more than 75 patents; however, his most revolutionary design is the single handle faucet. Issued in 1942, the single handle faucet became an instant success.
While still an engineering student at Washington State University, Moen worked nights at a garage to help pay for his tuition. One day, he burned his hands while washing them using a conventional two-handle faucet. From that experience, he resolved to create a faucet that would give the user water at a desired temperature. Moen designed a cylindrical valve with a piston action, we know this today as the Moen single handle cartridge. Moen's design was so popular that his faucet is in about 35 percent of homes today.

In designing the first ball valve in 1945, Landis H. Perry had specific objective in mind that was to create a combined volume and blended control valve. Perry also wanted to create a design that could be easily repaired. A patent was issued for Perry's ball valve in 1952. Shortly thereafter, Alex Manoogian purchased the rights to the patent and introduced the first Delta faucet in 1954. The Delta single-handle faucet was the first to use a ball-valve design and it proved very successful. By 1958, Delta's sales topped one million dollars.

Luther Haws was a self-employed plumber and sheet metal contractor in the early days of the 20th century. To those occupations, he added the job of sanitary inspector for the city of Berkeley, California. It was his role as an inspector that led to the invention of the first drinking fountain. In 1905, while on his rounds at a public school, Haws noticed the children drinking water out of a common tin cup that was chained to the faucet. He knew this was not a very sanitary situation, so when he returned to his plumbing shop he experimented with some spare parts. He showed his new design to the Berkley school department, which installed some of the first models.

Haws eventually gave up the plumbing company and formed the Haws sanitary Drinking Faucet Co. in Berkley in 1909. He received a patent on his product in 1911.

Watts Regulator Company introduced the first automatic temperature and pressure relief valve in the late 1930's. The new valve opened and closed automatically on both pressure and temperature. Prior to this invention, there had been only two forms of protection. One was the pressure only relief valve that did not control temperature, which was one cause of water heater explosions.

The second was a fusible plug-type temperature and pressure relief valve. When the temperature reached 210 degrees, the lead plug would melt. This created a discharge port and the overheated water flowed to atmosphere. The water continued to discharge until manually shut off. Watts discovery of the T&P valve has nearly eliminated water explosions and ensured that these appliances are safe for home use.

Invented by one of the original founders of Eljer in 1903, the vitreous china water closet cistern replaced the wall-hung wooden or copper- lined wooden cisterns used to flush a water closet. The vitreous china model received a great deal of resistance from the trade and the public.

The strength of the product was doubted so acceptance was slow. To prove just how sturdy china really was, Eljer hosted a demonstration. The cistern was laid on its back on a steel rail, a plank was placed on top of it, and 27 men stood on it. Eljer used the round cistern for approximately seven years. In 1908, the company introduced the first rectangular cistern that was cast in a mold – the method still used in the plumbing industry today.

Working with his own hand-made wood patterns in a vacant piano factory in Boston in 1939, Paul C. Symmons invented the first pressure balancing shower valve and founded the company that bears his name. Symmons discovered that sudden temperature changes in the shower water caused by pressure changes in the supply lines. Shower water got suddenly hot when someone somewhere else in the building turned on a cold-water faucet or flushed a toilet. Conversely, turning on the hot water faucet caused the shower water to turn cold.

Symmons perfected a valve that uses a hydraulic piston as the prime control units. As soon as the valve is turned on, both hot and cold water exert pressure on opposite ends of the piston, balancing it in the valve. If the hot water pressure drops, the piston reacts and reduces the cold inlet opening. The piston continually equalizes the pressure of hot and cold water, even when supply pressures change suddenly or drastically.

In 1896, Halsey W. Taylor lost his father to typhoid fever caused by a contaminated water supply. This personal tragedy led the young Taylor to dedicate his life to providing a safe, sanitary drink of water in public places. His quest led to the invention of the Double Bubbler, which projects two separate streams of water that converge to provide a fuller and more satisfying drink. The Double Bubbler also ensures that people drink far away from the actual projector. Taylor founded the company that bears his name in 1912. He developed his two-stream Mound building projector, as it was called then, for the U.S. government during World War I. Soon after, in 1926, he perfected and patented what is now known as the Double Bubbler.

Wisconsin architect John Hammes invented the food waste disposer in 1927. In his original model, the disposer ground up the food waste so it could be flushed down the kitchen drain. Hammes tinkered with his original model for 10 years before he launched the In-Sink-Erator Mfg. Co. and its line of disposers. When disposers were first introduced, most municipalities banned them because of worries about their impact on sewage treatment systems. However, by 1960 they were required in new construction by ordinance in more than 100 communities because of their sanitary value.

Much like the water closet, a disposer immediately removes raw food waste from the home through sewerage pipes to treatment plants, where it can be treated and neutralized. After more than 40 years on the market, disposers are now found in more than 50 percent of all households. About 4 million disposers are sold annually.

Elkay Mfg. Co. made a major breakthrough in 1974 with the invention of its non-pressurized water-cooling tank. Until then, most water cooler tanks were kept under pressure. Elkay engineers noted that with cold water under continual pressure within a cooler, a water leak or rupture could develop overnight or over a weekend, going undetected for many hours or days and causing extensive damage to the surroundings. In the new design, the only water under pressure was the water from the source up to the valve and the regulator cartridge. When activated, the push button valve opened to allow water to slow into the cooling tank and then out the bubbler. In the unlikely event of a burst tank, only the stored water within the cooling tank would be released.

In 1911, Kohler Co. introduced the industry's first one-piece recess bath with an integral apron. Before this time, built-in baths were cast in two separate sections-the tub proper and the apron. The apron and the tub were then fitted together by the plumber when either the tub was installed, or the two pieces were welded together at the factory before the fixture was enameled. The new one-piece tub, void of crevices, joins and seams, was much more sanitary and attractive than the two- piece forerunner. One of the most advanced products to come out of the 1920s was the Kohler electric sink. It was a combination of a conventional kitchen sink and the electric dishwasher almost as we know it today. The sink was an enameled iron fixture, massive and expensive, half sink, and half dishwasher. Invented by Kohler employee Frank Brotz in 1926, its only drawback was that it was a generation or two ahead of its time.

Kohler's introduction of bathroom sets (bathtub, toilet, and lavatory) in matching colors made fixtures much more than functional in 1927. Now instead of stark and sterile white, consumers could choose spring green, lavender, autumn brown, old ivory, or horizon blue.

For the first time, there was a concern for the aesthetics when planning a bath. However, beyond aesthetic considerations, the color concept was a revolutionary technological achievement in the plumbing industry of the time. The manufacturing of enameled cast iron fixtures and vitreous china fixtures required vastly different raw materials, processes, and techniques, so matching color was difficult.

William E. Sloan applied for a patent on his Royal Flush Valve June 13, 1906, and on Dec. 6, 1910, he was issued Patent Number 977,562.

With a flush valve installation, water flows under pressure from the supply piping directly to the fixture. Because of this connection to the water supply, they stand ready for repeated operations. In addition, because the water passes through the flush valve under pressure, the fixture is flushed with a scouring action to ensure proper cleansing. These two features, among others, are responsible for the popularity of flush valves in commercial buildings.

The first flush valves produced by Sloan Co. were designed for manual operation. Today, state of the art is flush valves are sensor-operated to make flushing automatic.

Through the 85 years that Sloan Valve Co. has manufactured the Royal flush valve, about 120 different companies have been in the same business at one time or another. Today, only a few remain. The first Americans awarded a patent for a water closet are James T. Henry and William Campbell. In 1857, their plunger closet resembled some of the twin-basin water closets developed and derided in England. These units were less than sanitary and shunned by some of the industry's earliest pioneers.

From the late 1850s to the mid-1890s, the number of patents granted for water closet designs grew as more and more inventors realized the potential market for an improved model. An American, John Randall Mann was granted a patent for his three-pipe siphonic closet in 1870. In 1876, William Smith earned his own for a jet siphon closet.

This model caught the attention of the famous American sanitary engineer George Waring who developed it into larger pieces of sanitary ware, as it was then called.

Thomas Kennedy, another American, improved on Mann's designed and patented a siphonic closet that required only two delivery pipes. One flushed the rim and the other started the siphon. Still further improvement occurred in 1890 with William Howell's water closet that eliminated the lower trap, but maintained the same superior function. By the turn of the century, water closet innovations were occurring on a nearly daily basis. The U.S. Patent Office received applications for 350 new water closet designs between 1900 and 1932. Two of the first granted in the new decade were to Charles Neff and Robert Frame.

These New Englanders were the first to produce a siphonic wash- down closet that would become the norm in this country in later years.

Fred Adee fixed problems with the bowl design in Neff and Frame's unit 10 years later. He redesigned the bowl, eliminating the messy overflows that sometimes occurred, and in doing so gave birth to production of the siphonic closet in America.

Some of the names of the other inventors who refined water closet design at this time have been lost, but their accomplishments have not. In the early 1900s, patents were granted for the flushometer valve, a backflow preventer, a wall-mounted closet with a blowout arrangement, a tank that rests on the bowl, and reverse trap toilets.

This is not to say there are not inventors alive and working today who will be added to this list of who's who in the years to come. However, some of the modern day water closet wonders are not plumbers or even plumbing engineers. They are scientists working on motors to create the "jet flush" toilet.

Engineers at the Emerson Motor Co. in St. Louis have developed a 3.3-inch motor and a 0.2 horsepower pump that fits in a toilet tank to add speed and power to each flush. These motorized toilets incorporate a steeper bowl than other gravity-style toilets to allow wastewater to flow out easier.

A slanted bowl and pressurized flush also allow the system to employ less water than a traditional gravity-flow toilet. To operate, the unit is plugged into a standard outlet in the bathroom. To date, Kohler Co. is the first plumbing manufacturer to market this technology.

Motors are affecting plumbing in other ways too. Emerson collaborated with pump manufacturers Zoeller Co. and Hydromatic Pump Co. to develop a plumbing system that liquefies waste. A pump is positioned in waste water pipes below the toilet and allows fixture manufacturers to meet existing water consumption requirements by chopping waste into a liquid consistency.

As waste moves through the system, a 5.5-inch, high-torque motor drives a sharp-tooth pump (much like that on a garbage disposer) that chops waste and toilet paper and pushes the resulting slush through the waste water system.

Types of Faucets

There are four basic faucet styles; in this section, we will look at each design type and basic repairs. Each manufacture, and there are several faucet manufactures out there, all use one of these four styles of mechanics. Once you have that figured out the repair, part is simple.

The four types of faucet designs are the ball, the cartridge, the compression, and the ceramic disc. The ball, disc, and cartridge are also known as washerless and for good reason, they do not have any washers to rely on for seating purposes. However, the compression style seals tightly against a brass seat at the base of the faucet body.

The first one we will look at is the compression style faucet. This is a common style in a well-known faucet brand called Chicago faucet.

Chicago Faucet Co. started out in 1901 by Albert Brown, as a plumbing fixture manufacture as well as lampshade frames, gas regulator valves, and oil burner tips.

By 1911, the demand for the company's plumbing products had grown to the point where it began marketing them under its own name and distributing its expanding line through wholesale plumbing supply houses.

In 1913, A.C. Brown invented a cartridge that laid the cornerstone of The Chicago Faucet Company. This major breakthrough in faucet design was the patented Quaturn cartridge. The replaceable, completely self-contained cartridge was revolutionary in its ability to turn water flow off from full flow with one-quarter turn of the handle.

Also unique was the way the cartridge closed with the flow of the water rather than against it, reducing washer wear and virtually eliminating drips. It was noteworthy that the cartridge was replaceable and interchangeable with other Chicago Faucets products. The Quaturn cartridge became the standard of reliability, durability, and value because of the Chicago Faucets commitment to standardization and renewability of parts.

The Quaturn cartridge has been updated over the years to incorporate new technology and materials, but it is still interchangeable with any Quaturn manufactured since 1913. In July 2002, the Geberit Group acquired Chicago Faucets. The Geberit Group, a 125 year-old company headquartered in Jona, Switzerland, is a European market leader and global provider in the area of plumbing technology. The company employs a staff of over 6,000 people worldwide.

The Chicago Faucets company employs over 500 people in operations located in Des Plaines; Milwaukee, Wisconsin; Michigan City, Indiana; and Elyria, Ohio. Chicago Faucets is one of only a few companies that still produce permanent mold, yellow brass castings right here in the US. In addition to standard castings, Chicago Faucets produces ECAST faucets, our line of durable, high-quality faucets, and fittings that are designed and manufactured with 0.25% or less total lead content by weighted average.

The compression faucet

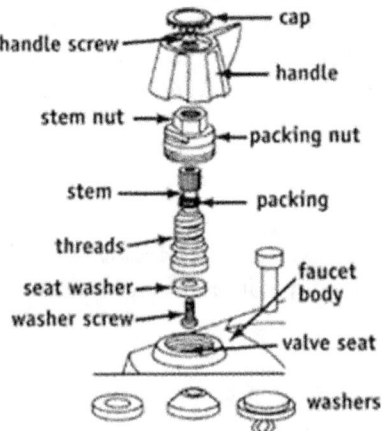

Fig. 1 Compression faucet

The compression faucet design has a simple mechanical operation. The handle is twisted, the stem moves up to open position raising the washer from the seat located in the faucet body. With that, simple motion water is allowed to flow from the body and directed to the spout.

Now for the repair work of the compression style faucet, it is just as easy as operating the faucet. First, remember to shut the water supply valve off first. Once the handle is removed, loosen the packing nut from the stem. Once the packing nut is removed, simply turn the stem to the left and it will thread right out of the faucet body. Now that the stem is out look at the bottom-side of the stem, you will notice a washer and a screw holding it in place. All you need to do is match up the washer with a new one and replace it. Inside the body of the faucet, you will see a brass seat, in order to remove the seat you will need a valve seat wrench.

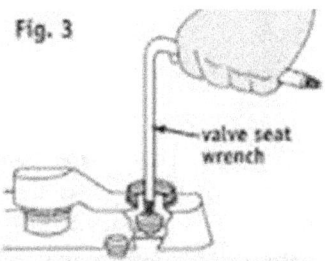

To remove the valve seat, insert a valve seat wrench into the faucet body and turn counterclockwise.

Tricks of the trade: this one faucet style can be repaired most of the time even without the need for parts. If the seat is rough and or jagged you can sand it smooth with sandpaper, also the sidewalk works well. If the seat is beyond sanding you will need to replace it as well. In addition, it is not unheard of to simple reverse the washer on the bottom of the stem. A few compression type faucets include hose bibb, or wall hydrant, mop sink faucets, commercial kitchen equipment, bathtub faucet, lavatory faucet, and kitchen faucets.

The cartridge faucet

Al Moen invented the best well-known cartridge style faucet. However, this style faucet is also manufactured in both single handle and two handle faucets.

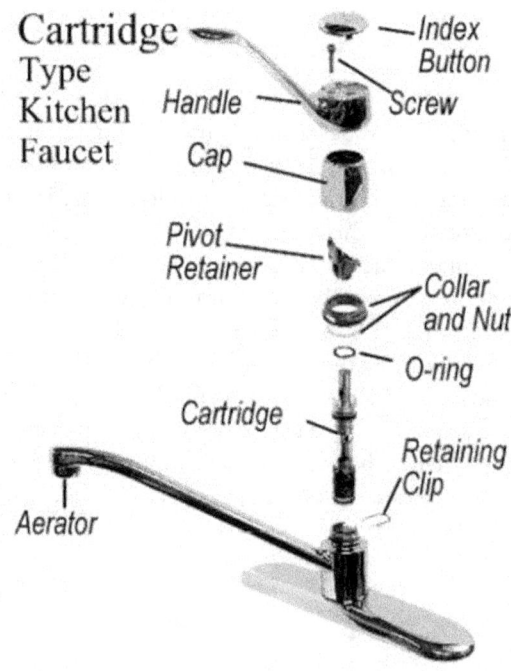

This style faucet is a simple faucet to repair, however, this faucet will require parts in order to make the necessary repairs. With that, you will also need a special tool in order to remove the older style cartridges. As with any faucet repair, be sure to turn the water supply off to the fixture before working on it. Secondly, you will need a screwdriver, a cartridge puller, and a pair of tongue and groove pliers or channel locks.

To begin remove the handle, cap, collar and nut. Secondly, remove the retaining clip (thin brass horseshoe clip). Most cartridge style faucets will have this holding the cartridge in place.

The next thing you will need is a cartridge puller as shown below.

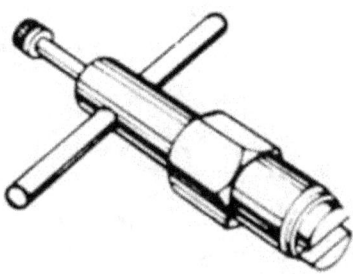

The cartridge puller is especially useful when pulling an older style cartridge such as the 1225B. Which are the much smaller version of the posi-temp common today.

Moen 1225B and retaining clip

The newer style posi-temp cartridge has a plastic pull sold in the package with the new cartridge. When removing an older 1225B style cartridge with the puller attached to the faucet, rotate the puller back and forth in order to loosen the cartridge from the faucet prior to pulling it.

Typically, if this is not done, you may end up pulling the center of the cartridge out and not the whole unit. If that should happen, not all is lost. You can use an extractor tool to remove the rest of the cartridge.

The Disc Type Faucet

The disc type faucet, pioneered by American Standard, is very similar to a cartridge faucet. One difference is the disc type faucet is much shorter and much wider than that of the cartridge style. In addition, there are usually three ports or openings on the underside of the disc style.

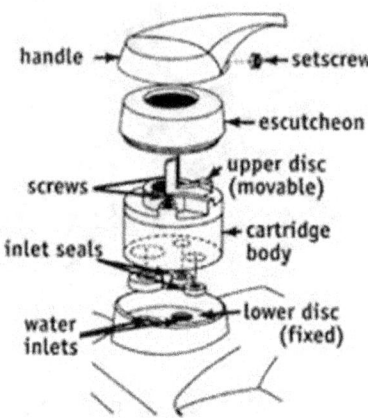

Ceramic disc faucets are the latest development in modern faucet technology. They are identifiable by their single lever over a wide cylindrical body. The disk faucet mixes hot and cold water inside a mixing chamber called a pressure balance cartridge.

Two ceramic disks at the bottom of the chamber will raise and lower to control the volume of water flow. Temperature can be controlled by a side-to-side rotation of the handle. The disc faucet mixes hot and cold water inside a mixing chamber called a pressure balance cartridge. Two ceramic discs at the bottom of the chamber will raise and lower to control the volume of water flow.

Ball Type Faucet

The ball style faucet is synonymous with the Delta faucet company, founded in 1954 by Alex Manoogian in Greensburg Indiana. This type of faucet requires just a small amount of patience when repairing. The handle of the ball-type faucet rests on a dome-shaped body and is attached by a setscrew. Use an Allen wrench to loosen the setscrew enough to lift off the handle. If the faucet is leaking from the base of the spout, use a spanner wrench (included in the ball-faucet repair kit) to tighten the locking collar by turning in a clockwise direction.

The repair kit for the ball-type contains the parts necessary to stop a leak from either source. (The ball mechanisms are usually sold separately, but these are seldom the source of the leak). If the leak stops once the locking collar has been tightened, no further repair is required. Just reattach the handle. If the leak continues, or is coming from the end of the spout, you will need to disassemble the faucet. First, close the water-shutoff valves under the sink. Use slip-joint pliers to twist off the domed cap. You may want to wrap the jaws of the pliers to avoid marring the chrome finish. Lift out the plastic cam and cam washer to expose the rotating ball.

Take out the ball and inspect it for signs of wear. Inside the faucet are two rubber valve seats that form a watertight seal against the rotating ball. Remove the valve seats by snaring them with a screwdriver. Use caution: there is a small spring behind each valve seat. If the valve seats appear worn, replace them by lining up new springs and seats on the end of a screwdriver and carefully dropping them into place. Use your finger to press them in firmly. New valve seats should stop any leaks coming from the end of the spout. If the leak originates from the base of the faucet, pull off the spout and inspect the O-rings. If they appear worn, pry them loose with the hooked end of the spanner wrench. Coat the new O-rings with heatproof plumbers grease, and pop them into place with the spanner wrench.

Reassemble the faucet by putting the parts back in this order: spout, ball, plastic cam and cam washer, and domed cap. Tighten the collar ring with the spanner wrench, and replace the handle. Turn the water on at the shutoff, and check for leaks.

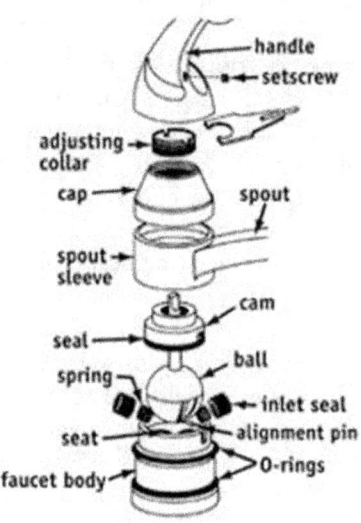

Delta ball type faucet

Water Closet

The water closet, by definition, is an enclosed room or compartment containing a toilet. As we learned in chapter one, the water closet was invented sometime around 1745-1755. We know it today simply as a toilet, commode, the porcelain throne, John, or a few colorful other names. In early America, a time before indoor plumbing, the frontiersmen had an outhouse. A small building constructed over an open pit, very similar to a modern day porta potty. The flush toilet did not gain popularity until after World War 1. When American troops came home from England full of talk about a "mighty slick invention called the crapper." The American slang term for the toilet, "the john," is said to be derived from the flushing water closets at Harvard University installed in 1735, and emblazoned with the manufacturer's name, Rev. Edward Johns.

The flush toilet is an invention of which humanity can be very proud. Without this marvelous contraption, disease would still be rampant and water supplies throughout the world would be undrinkable. The next time you see a toilet, remember the many inventors and plumbers that made it a clean, simple, easy-to-use device that makes our lives a little easier.

As with everything else in plumbing there are several toilet styles, manufactures we will look at the following, the gravity flush, the syphon jet, and the pressure assist. Before we get into the differences in the toilets let us look under the lid. In a modern toilet, we have two basic parts that make up the mechanics.

The first is called a fill valve, this is also known as an anti-syphon ballcock. Its name is derived from the little ball that acts as a float. This dates back around the 1940's. It was once made of aluminum and was threaded into a brass rod which in-turn was connected to the fill valve. The second part of the flushing mechanism is called the flapper. This is the rubber seal at the bottom of the tank; it is connected to the handle.

As the user flushes via the handle, the flapper will raise and allow water to flush, into the bowl. Among the variety of ballcocks available, the only style of importance is the anti-syphon ballcock. This device was patent around 1990, as people became more aware of the possibility of water contamination.

There became awareness of a water contamination through the toilet tank. After all the water within the toilet tank is the same water that comes from the kitchen sink faucet. The possibility is the water can be syphoned back from the toilet tank and into the water distribution piping.

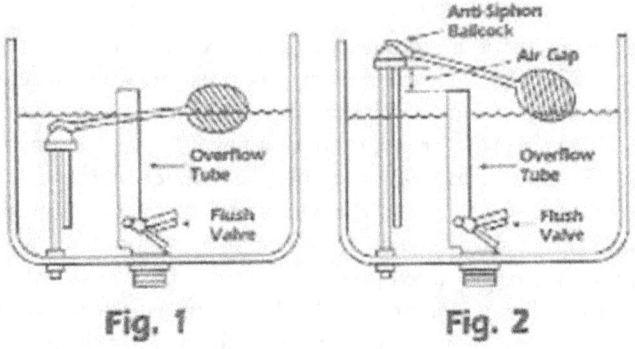

Fig. 1   Fig. 2

This can occur if the water level in the tank is above the fill valve. As depicted in the diagram above. The anti-syphon ballcock is raised above the water level creating an air gap, between the water supply into the toilet tank. It is important to point out the water level in the tank will travel or flow down the overflow tube in the event the fill valve should malfunction. This will ensure the toilet tank water never flow to the point of the inlet portion of the fill valve.

I will always tell either an apprentice or homeowner it is best practice when servicing a flush style water closet to replace both the fill valve and flapper at the same time. The parts will wear the same and if one is requiring a replacement, the chance of the other needing service is a matter of days or weeks in most cases.

Types of Water Closets

We will look at several designs. However, the first water closet we will focus on is the gravity toilet, this is the style most people are familiar with, it is also the "builder grade" or least expensive. The gravity flush toilet design depends upon the volume and weight of the water in the tank to create the flushing action. This can be summed up by stating the toilet works on principals of gravity, hence the name.

Water housed in a tank above the bowl is released into the bowl, causing water to push through the trap, thus emptying the bowl of its contents. One last thing to point out is there will be only an outlet in the bottom of the bowl, rather than two as with a syphon-jet style.

Low-Flow

In the U.S. and some Canadian locations, the maximum flush volume of a toilet is set at 1.6 gallons. These toilets are called "low-flow", "ultra-low-flow" or some similar term. Up until now, the 1.6-gallon toilet has dominated the marketplace.

High-Efficiency (HET)
Beginning in 1999, however, manufacturers began introducing what are known as "high-efficiency" toilets, or HETs. HETs are defined as having an effective flush volume of 1.28 gallons or less. The HET category includes both single flush and dual-flush fixtures. The current database contains 2,680 different HET models certified to the U.S. EPA Water Sense specification. California, Georgia, Colorado, and Texas currently prohibit installation of non-HET fixtures (with some exceptions).

Single Flush
Single flush HETs are available that flush as little as 0.8 gallons and as much as 1.28 gallons. Today, there are 108 toilet brands offering **1,647** models of single-flush.

Dual-Flush
Dual-flush HETs provide the user with the option of a "full" flush for normal operations or a "reduced" flush for liquids only. They are subject by the national codes and standards to a maximum flush volume of 1.6 gallons in the "full" flush mode and 1.1 gallons in the "reduced" flush mode. (Most dual-flush toilets, however, use about 0.8 gallons in the reduced flush mode.) Dual-flush toilets installed in residences are flushed in the reduced mode once for every full flush.

The average flush volume for a typical dual-flush toilet, then, is less than 1.28 gallons, thereby meeting the requirements for an HET. Today, 93 brands of dual-flush toilets offer 900 different dual-flush models in the North American marketplace.

Siphonic
In most toilets sold in North America, the force of the water coming from the tank acts to create a siphon in the exit from the bowl (known as the "trap-way"). As a result, the siphon action pulls the waste through the trap and into the drain. Once this happens, an automatic valve begins to refill the emptied tank, getting the toilet ready for the next flush.

Wash-Down
Less common in North America is the wash-down design of the toilet bowl. In these fixtures, no siphon is created in the bowl trap-way. Rather, the weight of the water simply pushes the waste into the trap-way. Wash-down toilets, therefore, usually feature a much larger diameter trap-way (since no siphon is required) and may result in less clogging than a siphonic design. As with siphonic toilets, the tank begins to refill once the flush is completed.

The disadvantage of the wash-down bowl design is the small water surface area in the bowl (the 'water spot').

North Americans are not accustomed to small water spots and find that this smaller bowl results in streaking. On the other hand, wash-down bowls seldom clog (because the trap-way is twice the cross-sectional area of a typical siphonic bowl).

Pressure-Assist
Pressure-assist toilets use compressed air to create a more forceful flush action. These toilets take advantage of the pressure of the building water supply to get the job done, with the help of a pressure containment vessel. Water from the supply line is forced into the air- filled pressure tank at the building's water pressure of 40 to 80 psi. The air in the tank is compressed and when the flush lever is pushed, the water rushes into the bowl.
One benefit of a pressure-assist toilet in humid climates is that the water is contained inside the pressure vessel, which, in turn, is inside the vitreous china toilet tank. That design results in little or no tank sweating.
Because of the design of the pressure-assist toilet, the flush action may be somewhat louder than a gravity-fed toilet. However, that sound is of very short duration, unlike a gravity-fed toilet that can take 20 to 30 seconds to refill.

Round Front vs. Elongated Bowl Design
Today's toilets come with either a smaller, round-front bowl or a longer, elongated-front bowl. The round-front bowl is ideal for compact bathroom spaces and is found extensively in older homes. Elongated bowls with a longer rim dimension are more comfortable for adult use and improved hygiene. Plumbing codes require elongated bowls in commercial applications.

Conventional vs. Elevated-Height Bowls (ADA Height)
Again, consumers and designers have choices when it comes to bowl dimensions. Two bowl heights are available: the conventional height bowl with a top rim generally in the range of 14 to 15 inches from the floor (excluding any toilet seat) and the chair-height bowl ranging from 16.5 to 18 inches. The Americans with Disabilities Act of 1990 sets the minimum bowl height at a 17-inch minimum. This chair-height design is becoming increasingly popular among all ages of the population and is often referred to as an "ADA height" bowl. Manufacturers have chosen their own terminology, including such descriptors as "comfort height", "right height", or some similar term.
Our old 5-gallon toilets could flush cigars, feminine-hygiene products, bullets, Popsicle sticks and an occasional diaper with little effort. The new 1.6-gallon toilets sometimes struggle to flush our humble daily loads of human waste. However, the law is the law, as those who favor the law are fond of saying. Moreover, the law is the National Energy Policy Act of 1992, which requires that all toilets made or sold in this country meet new federal water-efficiency standards.

To conserve water, those standards set the upper limit of a single flush at 1.6-gal. The law took effect in 1994 for residential toilets and in 1997 for commercial toilets. Complaints against the 1.6-gallon toilet include sluggish or incomplete flushing; a small "water spot," as the area of the toilet bowl water surface is called; staining; and the need to double- flush or triple-flush. Critics say that if a 1.6-gallon toilet is flushed more than twice it uses more water than the now illegal 3.5-gallon toilets.

A major toilet manufacturer apologizing for the fact that their 1.6- gallon toilet did not work well but that they were required to make it. Sometimes the problem is in the pipes. A number of plumbers warned that installing a 1.6-gal Gravity-flush toilet in an old house could lead to clogs and backups.

Often, most waste pipes are 4-inch or greater diameter cast iron, which is a lot rougher on the inside than modern plastic pipe. When the cast- iron pipe was installed, toilets flushed anywhere from 5 gallon to more than 7 gallons of water. Now that they are down to 1.6-gallons, that is often not enough water to power the waste through.

Houses that have 4-in. to 6-in. cast-iron drains are a problem, when you put a 1.6-Gal. toilet in with that diameter pipe; it just barely makes the bottom of the pipe wet. As a retrofit in a house with old plumbing, it is lousy. Now I run high-use fixtures, like the washing machine, just after the toilet. The washing machine will help move that waste down the line.

We are using plumbing fixtures that were designed for the 1990s and putting them in plumbing systems that were designed for the 1920s.

The Bidet

A bidet shall be equipped with hot and cold, tempered and cold, or tempered water only. An atmospheric vacuum breaker shall be installed at the discharge side of the flushing valve. The bottom of the vacuum breaker, or critical level line shown on the vacuum breaker, shall be at least four inches above the overflow rim of the bidet. The French invented the use of the bidet and it was very crude in the beginning. It was just a bowl that you squat over and then evolved into a separate porcelain toilet with faucets, but you still had to squat or straddle it.

In the middle 1900's American toilet manufacturers improved the bidet as a porcelain bathroom device, but the American culture did not adopt it and it was mostly exported to other countries. The bidet is actually a small horse. In addition, you straddle a horse, so when the French used a device for cleaning the genitals and the posterior they called it a bidet, because you straddle it when you wash. The earliest known written reference to the bidet is dated 1710. In 1750, the bidet à seringue appeared. It provided an upward spray with a hand-pump fed by a reservoir. Until the 1900, the bidet was confined to the bedroom, along with the chamber pot (a bucket that served as a toilet.) Modern plumbing brought the bidet into the bathroom, where it sits next to the toilet. It is resembles a toilet, but it has facet knobs and you straddle it to wash your private parts, after using the toilet.

Today, there are many bidets (modern porcelain toilet bowls) to choose from built by many different manufacturers, but they take up extra bathroom space. However, another device has evolved that is attached to an existing toilet. This is more of an efficient use of bathroom space. The evolution of the bidet started after WWII, when the Japanese started importing American made toilets. It was better than Japanese toilets, because you could sit down in comfort, rather than squatting over old Japan toilets.

After the adoption of American made toilets, including American made bidets (porcelain toilets that required you to squat), the Japanese decided to improve the product. The Japanese liked the style of the American toilet that allowed you to sit and the Japanese invented a device that you attach to your existing toilet, shoots water, through a jet valve, and cleans the posterior areas without the use of toilet paper. This ingenious manual device was then expanded to include hot and cold water. Electronic models were then designed to include a heated seat, retractable cleaning jets, sensors, automatic controls, a dryer, and a deodorizer. Bidets are used by many European and Asian countries and have not yet penetrated the American marketplace. In continental Europe, the usefulness of the bidet is fully understood and is considered as important in the bathroom as the toilet and the tub – no well-equipped home is without one.

However, most Americans have never seen a bidet. Those who have, generally observed them in upscale hotels, either in the U.S. or in Europe. Rare is the American home that actually has one.

To some, this seems a bit strange, considering the American preoccupation with cleanliness. However, the majority of Americans start their day in the shower, rather than visit the bathtub once a week. Thus, the use of the bidet for personal hygiene has not yet taken on an important role in America.

The Urinal

The urinal was first patented in the United States immediately following the Civil War, when Andrew Rankin introduced an upright flushing apparatus in 1866. The device enjoyed widespread popularity in large northern cities. A sanitary plumbing fixture for urination only, predominantly used by males. It can take the form of a container or simply a wall, with drainage and automatic or manual flushing, or without flush water as is the case for waterless urinals. Most public urinals incorporate a water flushing system to rinse urine from the bowl of the device, to prevent foul odors. One of several methods can trigger the flush.

Manual Handles

This type of flush might be regarded as standard in the United States. Each urinal is equipped with a button or short lever to activate the flush, with users expected to operate it as they leave. Such a directly controlled system is the most efficient, if patrons remember to use it.

This is far from certain, however, often because of fear of touching the handle, which is located too high to kick. Urinals with foot-activated flushing systems are sometimes found in high-traffic areas; these systems have a button set into the floor or a pedal on the wall at ankle height.

The Americans with Disabilities Act requires that flush valves be mounted no higher than 44 inches AFF (above the finished floor). Additionally, the urinal is to be mounted no higher than 17 inches above the finish floor, and to have a rim that is tapered and elongated and protrudes at least 14 inches from the wall. This enables users in wheelchairs to straddle the lip of the urinal and urinate without having to "arc" the flow of urine upwards. Some urinals are equipped with water-saving "dual-flush" handles, which use half as much water when pushed upwards, and operate a standard full flush when pressed downwards. The handles are often color-coded green to alert users to this feature.

Timed Flush

In Germany, the United Kingdom, France, the Republic of Ireland, Hong Kong and some parts of Sweden and Finland, manual flush handles are unusual. Instead, the traditional system is a timed flush that operates automatically at regular intervals. Groups of up to ten or more urinals can be connected to a single overhead cistern, which contains the timing mechanism. A constant drip feed of water slowly fills the cistern until a tipping point is reached, when the valve opens (or a siphon begins to drain the cistern), and all the urinals in the group are flushed. Electronic controllers performing the same function are also used.

This system does not require any action from its users, but it is wasteful of water when toilets are used irregularly. However, in these countries users are so used to the automatic system, that attempts to install manual flushes to save water are generally unsuccessful. Users ignore them not through deliberate laziness or fear of infection, but because activating the flush is not habitual. To help reduce water usage when restrooms are closed, some restrooms with timed flushing use an electric water valve connected to the restroom light switch. When the building is in active use during the day and the lights are on, the timed flush operates normally. At night when the building is closed, the lights are turned off and the flushing action stops.

Automatic Flush

Electronic automatic flushes solve the problems of previous approaches, and are common in new installations. A passive infrared sensor identifies when the urinal has been used, by detecting when someone has stood in front of it and moved away, and then activates the flush. There usually is also a small override button, to allow optional manual flushing.

Automatic flush facilities can be retrofitted to existing systems. The handle-operated valves of a manual system can be replaced with a suitably designed self-contained electronic valve, often battery-powered to avoid the need to add cables. Older timed-flush installations may add a device that regulates the water flow to the cistern according to the overall activity detected in the room. This does not provide true per-fixture automatic flushing, but is simple and cheap to add because only one device is required for the whole system.

To prevent false triggering of the automatic flush, most infrared detectors require that a presence be detected for at least five seconds, such as when a person is standing in front of it. This prevents a whole line of automatic flush units from triggering in succession if someone just walks past them. The automatic flush mechanism also typically waits for the presence to go out of sensor range before flushing. This reduces water usage, compared to a sensor that would trigger a continuous flushing action the whole time that a presence is detected.

Waterless Urinals

Since about the 1990s urinals are available on the market that use no water at all. These are called waterless urinals or flushless urinals. A German named Klaus Reichardt, who secured his innovation with several patents, first invented them. Waterless urinals can save between 15,000 and 45,000 US gallons (57,000 and 170,000 l) of water per urinal per year, depending on the amount of water used in the water-flushed urinal for comparison purposes, and the number of uses per day. For example, these numbers assume that the urinal would be used between 40 and 120 times per business day. Models of waterless urinals introduced by the Waterless Company in 1991 and others in 2001. Falcon Water free Technologies and Sloan Valve Company, as well as Duravit, use a trap insert filled with a sealant liquid instead of water. The lighter-than-water sealant floats on top of the urine collected in the U-bend, preventing odors from being released into the air.

The cartridge and sealant must be periodically replaced. Waterless urinals may also use an outlet system that traps the odor, preventing the smell often present in toilet blocks, another method to eliminate odor was introduced by Caroma Co. that installed a deodorizing block in their waterless urinal that was activated during use. The drainpipes from waterless urinals need to be installed correctly in terms of diameter, slope, and pipe materials in order to prevent buildup of calcium phosphate precipitates in the pipes, which would cause blockages and could require expensive repairs. In addition, the undiluted urine is corrosive to metals (except for stainless steel), which is why plastic pipes are generally preferred for urine drainpipes. US federal law has mandated no more than one gallon per flush since 1994, and the EPA estimates that the average urinal is flushed 20 times per day, which gives an average water use of 7,300 gallons (28,000) per year. Mechanical traps are not allowed by US building codes but are allowed in many other countries.

Plumbers and code officials initially opposed waterless urinals, citing concerns about health and safety, which have been debunked by scientists who have studied the devices. Facing opposition to their attempts to have the devices allowed in plumbing codes, manufacturers devised a compromise.

The Uniform Plumbing Code was modified to allow waterless urinals to be installed, if unneeded water lines were nevertheless run to the back of the urinals. This allows conventional water-flushing urinals to be retrofitted later, if waterless models were judged unsatisfactory over time. This style urinal is however approved in Illinois for use.

Bathtubs & Whirlpools

Most modern bathtubs are made of acrylic or fiberglass, but alternatives are available in enamel on steel or cast iron. The end of World War I brought with it a construction boom in the US. Bathrooms were fitted with a toilet, sink, and bathtub – mostly claw-foot bathtubs. Nevertheless, in 1921, only one percent of homes in the US had indoor plumbing. Outhouses were still the norm in rural America. The Sears catalog, with its uncoated, absorbent pages, was a popular form of toilet paper often found hanging inside the outhouse.

Over time, the once popular claw-foot tub morphed into a built-in tub with apron front. This enclosed style afforded much easier maintenance of the bathroom and with the emergence of colored sanitary ware, more design options for the homeowner. Crane Company introduced colored bathroom fixtures to the US market in 1928.

The trend today, though, is shifting back to the elegant style and luxury of a soaking claw-foot tub. Homeowners are tearing out their dime-a-dozen built-in tubs and replacing them with reproduction roll rim footed tubs. Now available in both the classic cast iron and lighter weight acrylic styles, claw-foot bathtubs are produced in a variety of styles and foot finish options.

Modern bathtubs have overflow and waste drains and may have taps mounted on them. They are usually built-in, but may be freestanding or sometimes sunken. Until recently, most bathtubs were roughly rectangular but with the advent of acrylic thermoformed baths, more shapes are becoming available. Bathtubs are commonly white in color although many other colors can be found. The process for enameling cast iron bathtubs had been invented by the Scottish-born American David Dunbar Buick. In the 1880s while working for the Alexander Manufacturing Company in Detroit.

The company, as well as others including Kohler Company and J. L. Mott Iron Works, began successfully marketing porcelain enameled cast-iron bathtubs, a process that remains broadly the same to this day. Far from the ornate feet and luxury most associated with claw-foot tubs, an early Kohler example was advertised as a "horse trough/hog scalder, when furnished with four legs will serve as a bathtub." The item's use as hog scalder was considered a more important marketing point than its ability to function as a bathtub.

Two main styles of bathtub are common:

- Western style bathtubs in which the bather lies down. These baths are typically shallow and long.
- Eastern style bathtubs in which the bather sits-up. These are known as ofuro in Japan and are typically short and deep.

Whirlpools

Whirlpool tubs first became popular in America during the 1960s and 70s. A spa or hot tub is also called a "Jacuzzi" since the word became a generic after plumbing component manufacturer Jacuzzi introduced the "Spa Whirlpool" in 1968. Air bubbles may be introduced into the nozzles via an air-bleed venturi pump. Founded in 1915 by seven brothers, led by Giocondo Jacuzzi, Jacuzzi and Brothers made wooden propellers under military contracts, based at 2043 San Pablo Ave, Berkeley, California. In 1920, the brothers also dabbled briefly with aircraft design and manufacture, with a single-seat monoplane and a seven-seat cabin monoplane. Both aircraft were noted for their use of laminated wood products for fuselage manufacture, but were essentially unsuccessful, with only one of each type being built. By 1923, the company was styled as Jacuzzi Bros Propellers, with headquarters at 1450 San Pablo Ave. The company survived to become famous as the whirlpool and bubble bath manufacturer of today, with the name Jacuzzi becoming synonymous with the bath products regardless of manufacturer.
In October 2006, Apollo Management, a large private equity firm, announced a $990 million leveraged buyout of Jacuzzi Brands. In 2008, Jacuzzi moved its world headquarters to The Shoppes at Chino Hills, California. In May 2012, Jacuzzi Group Worldwide acquired the assets of ThermoSpas, a Wallingford, Connecticut-based manufacturer of spas and hot tubs.

Mixing Valves

Plumbing code note, the following regarding water temperature settings for the following fixtures, all-public hand sinks must not exceed a water temperature of 110 degrees. Tub/shower or combination thereof must not exceed a water temperature of 115 degrees. These plumbing fixtures must be installed with temperature regulating equipment in order to prevent the risk of scald.
The tempering valve as shown in *fig.2* is commonly used to temper water for a single use public hand sink. This valve allows hot and cold water to mix, and can be adjusted by means of an adjustment screw. In some valves there is a handle with a vandal proof set screw for adjustment. This is to prevent an unauthorized person from tampering with the valve.

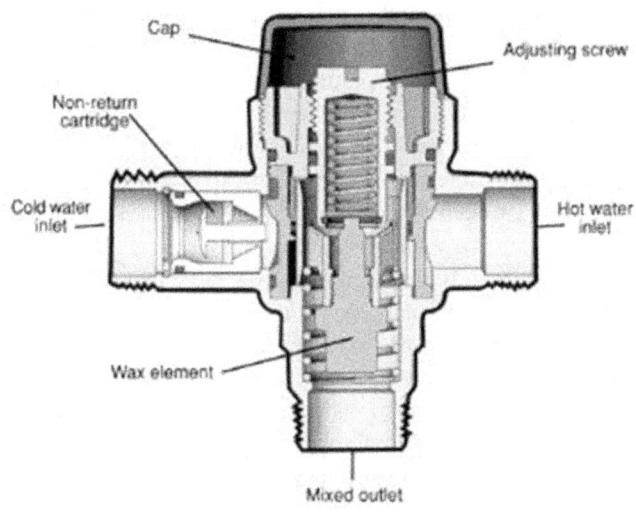

Tempering valve Figure 2

There are also tempering valve or mixing valve as they are commonly known for use with multiple use or a "bank" of public hand sinks.
They can be adjusted to a not-to exceed or pre-determined temperature. Often this type of valve is installed within the supply or feed piping to the hand sinks, usually in the ceiling with an access panel. All tub and shower valves manufactured after 1990 had been required to meet the "new" federal anti-scald protection. This had actually begun back in the early 1970s, with studies conducted by the American Society of Sanitary Engineering (ASSE). Symmons Industries Inc. pioneered the pressure balancing shower valve to address the thermal shock issue. The early valves provided thermal shock protection but not anti-scald protection. In later years, manufacturers added setscrews, limit stops, and cams to allow the valves to be set to limit the amount of rotation toward the hot water, thereby providing a safe temperature for showering or bathing. The severity and risk associated with scalding increases drastically with relatively small increases in water temperature. An anti-scald shower valve will only protect a bather if the maximum temperature limit stop or setscrew is adjusted when the valve is installed.
It is very important that the plumber installing the valve to make sure the water heater is set to the proper temperature and water is flowed to the tub/shower valve in order to set the maximum temperature limit stop to a maximum of 115 degrees.

In Illinois, the water heater temperature control is not an acceptable means by which temperature can be controlled. While working as an inspector that has come up once or twice. I have had contractors and consumers say the same thing. "How about I just dial the water heater down." The answer is no, under 890.690 (B) found in the (F section, plumbing fixtures of the codebook.) Federal plumbing specification was the minimum standard for plumbing products used by the federal government. The federal specifications were the first to mandate single-handled shower mixing valves for federal government projects. Early valves came in three categories: type P, pressure balancing, and type T, thermostatic, and type M, mechanical mixers.

The mechanical mixer is simply a single-handed valve that opened with cold water and added hot water as the valve rotated open, as with the operation of all tub/shower valves available on the market today.

These type of valves have evolved to include type P, type T, and type TP (combination thermostatic/pressure balancing) valves. Type P devices have a pressure balancing piston or diaphragm that equalizes the pressure between the two inlet ports and maintains the same outlet temperature, as long as the incoming temperatures remain the same.

Type T devices have a temperature-sensing element that adjusts the inlet ports to maintain a relatively constant outlet temperature. Type TP devices have pressure-sensing and temperature-sensing elements and can compensate for both temperature and pressure changes.

The scald burn studies done by doctors showed that it took approximately eight minutes of exposure to temperatures in the range of 120 F for adults to develop a serious scald burn: Someone exposed to water at 120 F would have up to eight minutes to get out of harm's way before an injury started to develop. Children, infants, and the elderly have skin that is thinner and could develop burns sooner than eight minutes.

The 120-degree temperature limit has become an industry standard for scald prevention in showers and combination bathtub/showers. Again, however in Illinois, it is important to remember the maximum setting for and combination bathtub/shower is 115 degrees.

Table 1 – Water Temperature Effects on Adult Skin
(Source: Report prepared by Dr. Moritz and Dr. Henriques at Harvard Medical School in the 1940s)

Temperature	First degree burn	Second degree burn
111 degrees	270 minutes	300 minutes
113 degrees	120 minutes	180 minutes
116 degrees	20 minutes	45 minutes
118 degrees	15 minutes	20 minutes
120 degrees	8 minutes	10 minutes
140 degrees	3 seconds	5 seconds
150 degrees	instant	2 seconds

Dishwashing Machines

Domestic dishwashers must have the discharge of the machine piped to a trap from under the kitchen sink. They may not discharge into the "knock-out" of the disposal. The discharge must also be looped to the underside of the overflow rim of the sink. This is an important note because homeowners, and or the store in which they have purchased the dishwasher from, will attempt to connect the unit without realizing if it is installed via the garbage disposal, food rotting and bacteria from under the blade mechanism will develop a foul odor. When the sink should back-up the rotting food and bacteria will spill or flow back into the dishwasher.

Emergency Eye Wash/Shower

Emergency eyewash and shower units are designed to deliver water to rinse contaminants from a user's eyes, face, or body. As such, they are a form of first aid equipment to be used in the event of an accident.
However, they are not a substitute for primary protective devices (including eye and face protection and protective clothing) or for safe procedures for handling hazardous materials. The selection of emergency eyewash and shower equipment is often a complicated process. In addition to addressing design and engineering issues, specifiers must be aware of regulatory requirements and compliance standards. A common reference point when selecting emergency equipment is ANSI/ISEA Z358.1, "Emergency Eyewash, and Shower Equipment." This standard is a widely accepted guideline for the proper selection, installation, operation, and maintenance of emergency equipment.

The Occupational Safety and Health Act of 1970 was enacted to assure that workers are provided with "safe and healthful working conditions." Under this law, the Occupational Safety and Health Administration (OSHA) was created and authorized to adopt safety standards and regulations to fulfill the mandate of improving worker safety.
OSHA has adopted several regulations that refer to the use of emergency eyewash and shower equipment. The primary regulation is contained in 29 CFR 1910.151, which requires where the eyes or body of any person may be exposed to injurious corrosive materials, suitable facilities for quick drenching or flushing of the eyes and body shall be provided within the work area for immediate emergency use.

The 2014 version of the standard states that the water temperature delivered by emergency equipment should be 'tepid.' Tepid is defined to be between 60°F (16°C) and 100°F (38°C). However, in circumstances where a chemical reaction is accelerated by flushing fluid temperature, a facilities safety/health advisor should be consulted to determine the optimum water temperature for each application. The delivery of tepid water to emergency equipment may raise complicated engineering issues. At a minimum, it generally involves providing both hot and cold water to the unit and then installing a mixing valve to blend the water to the desired temperature. In general, the ANSI standard provides that emergency equipment be installed within 10 seconds walking time from the location of a hazard (approximately 55 feet.) The equipment must be installed on the same level as the hazard (i.e. accessing the equipment should not require going up or down stairs or ramps).

The path of travel from the hazard to the equipment should be free of obstructions and as straight as possible.

Plumbed emergency equipment must be connected to a potable water supply line. It may be advisable to install a shut off valve on the water line, upstream of the unit, to facilitate maintenance of the equipment.

If a shut off valve is installed, provision must be made to prevent unauthorized closure of the shut off valve. Once connected to a water supply line, water will enter the emergency equipment and stand in the unit up to the valve(s). When activated, water will flow through the entire unit. Therefore, the unit must be constructed of materials that will not corrode when exposed to water for extended periods. Materials that are considered acceptable for this purpose include brass, galvanized steel, and many types of plastics (ABS, nylon, etc.). However, these materials may not provide durable service when exposed to harsh industrial conditions may deteriorate in direct sunlight or be subject to other limitations. After an emergency eyewash or shower unit has been used, the wastewater may contain hazardous materials that cannot or should not be introduced into a sanitary sewer. It may be necessary to connect the drain piping from the emergency equipment or floor drain to the building's acid waste disposal system or to a neutralizing tank.

Section 890.800 (D) A request for permission to install special plumbing fixtures or items designed for a particular purpose requiring water and waste connections not otherwise provided for in this part shall be submitted to the department for approval prior to installation.

With the installation and design of a plumbing system, for any building that will be designated as a public building there are requirements for the plumbing fixtures to consider. The first is regarding the building use. Is the building a restaurant? On the other hand, public school?

What is the full capacity of the building? All these questions come into the design part of a new building.

Then consider the ADA code, and the possibility of the use of "green plumbing". During the planning stage, these things collectively must be a focus.

The architect and possibly a plumbing engineer will draw plans and design these elements, and there will be a plan review as well. Nevertheless, just because other individuals are involved does not mean the plan is set in concrete so to speak. It is also the job and responsibility of the plumber, supervisor, or superintendent to make sure all of the necessary guidelines are correct during construction. Whenever a new project starts, it is a wise decision to become familiar with the drawings and fixture schedules.

Large corporations that have many chains of restaurants, for example, across America are have very specific plans right down to a certain model toilet seat. The restaurant will more than likely have someone to come in prior to the grand opening to verify that the fixtures that have been specified on the plans are exactly what has been installed. If the exact model numbers are not the same you will be replacing items before the opening of that store.

I have been involved with such construction projects and in a pre- construction, meeting all contractors present where told just that. Follow the drawings exact.

9 HANGERS AND SUPPORTS

Types of Support Systems

From the beginning, there has always been a way to support piping and the weight of whatever is to be carried through it. For example, look at the Roman Aqueducts, constructed around 2700 B.C.
The Aqueducts not only served to carry the precious water to the city, but acted as the support as well. Lucky for the modern day plumber, supports and hangers have come a long way. Gone are the days of using heavy blocks of stone and the need to build giant arches to carry the weight of the liquid.
Section (G) 890.910 states that, in general, hanger's anchors and supports must be of material and strength to support piping and its contents. That is straightforward- however, this very important step in the installation of plumbing is often taken for granted.
Most often, the use of inappropriate hangers- whether that may be the wrong material or a piece of wire wrapped around a 4-inch sewer line- is reason to fail a plumbing inspection. In the same manner, I have seen the use of very expensive hangers used to support a vertical ½- inch copper pipe in a wall. During the planning stage, it is as important as the pipe methods and should be considered a major part of any plumbing installation. All too often, I have had plumbing contractors completely forget to include any type of hangers, or support system into an estimate. This usually means that there is a lot of money that the contractor is going to give away.
A few of the basic hangers and support systems such as "j-hooks" for plastic, gas, or copper can become very costly if not included in the bid. These materials as well as solvent cement, primer, solder and flux all need to be considered prior to submitting a bid.

Materials and Support Types

The trapeze style support is exactly what you are thinking. This is used for long runs of and capable of holding multiple piping systems. This system is often used in a parking garage or similar type structure. It is held into the ceiling usually concrete, with lead anchors and all thread.

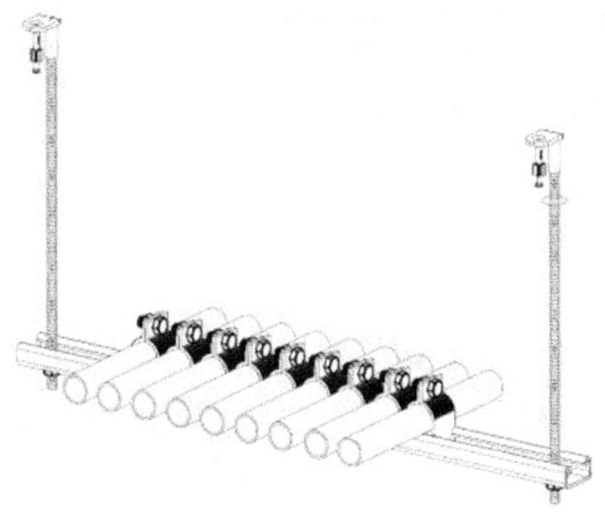

This method is most often fabricated on the job site. It takes some labor and time to build this system however; it is extremely effective if you have multiple piping systems to install within the same area of a building. Again using the right support system for the application is of importance. This type of support would be most effective in a commercial building.

Another type of support used in commercial is the loop hanger, or pear hanger as it is also known, is designed for a single pipe support. They are available in a wide selection of finishes and sizes.

LOOP HANGER

Material: Steel
Finished: zinc plated, epoxy coated, copper plated

SIZE	ZINC PLATED CAT#	EPOXY COATED CAT#	COPPER PLATED CAT#
1/2"	ALX-750Z	ALX-750E	ALX-750C
3/4"	ALX-751Z	ALX-751E	ALX-751C
1"	ALX-752Z	ALX-752E	ALX-752C
1-1/4"	ALX-753Z	ALX-753E	ALX-753C
1-1/2"	ALX-754Z	ALX-754E	ALX-754C
2"	ALX-755Z	ALX-755E	ALX-755C
2-1/2"	ALX-756Z	ALX-756E	ALX-756C
3"	ALX-757Z	ALX-757E	ALX-757C
3-1/2"	ALX-758Z	ALX-758E	ALX-758C
4"	ALX-759Z	ALX-759E	ALX-759C
5"	ALX-760Z	ALX-760E	ALX-760C
6"	ALX-761Z	ALX-761E	ALX-761C
8"	ALX-762Z	ALX-762E	ALX-762C

With this type of hanger, it is important to note if you are following the 2015 energy code, the hanger size will need to be large enough in order to accommodate pipe insulation. Furthermore, this type of hanger will require the use of all thread as shown below. This also is available in a variety of sizes and most common sold in lengths of ten feet.

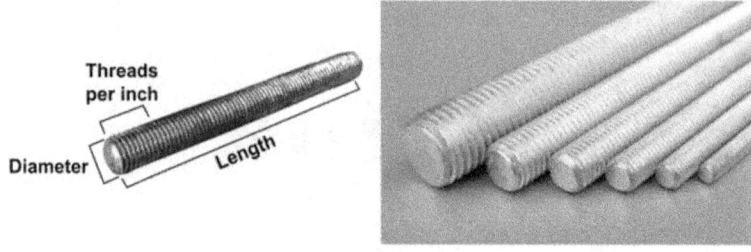

All-thread

The all-thread can be cut down to whatever the desired length requirements for elevation. Once again, this type of support is common in commercial applications. With the use of all-thread, it is also common practice to use beam clamps to fasten securely to the building. Another code advisory is to be sure to attach the beam clamp to the top of a steel truss.

Beam Clamp

Clevis hanger is very similar to the loop hanger it is slightly more expensive and is able to carry much more weight than a loop hanger.

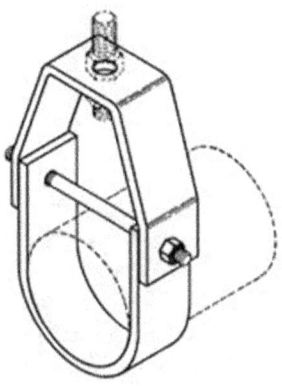

Clevis Hanger

The types of supports and hangers we have looked at thus far we have been focused around the commercial construction applications. Let us look at a few specifics in a residential application, and how a few of these hangers can be utilized. The PVC "j-hook" is one of the most common means to support a PVC sewer piping system in a basement.

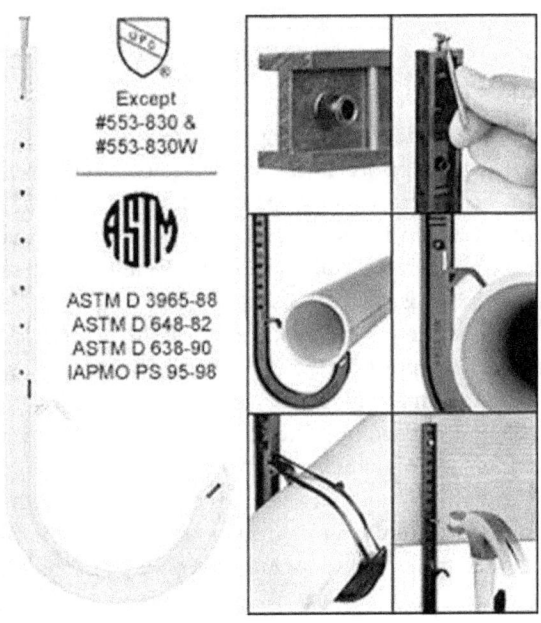

When supporting copper tube there are a number of hangers available as well these should be of the same material as the pipe being supported. In other words, you do not want to install a galvanized support while installing copper.

Wire Hook

Another common way to properly support copper in a residential application is with the use of a Sioux Chief Brand pipe nailer.

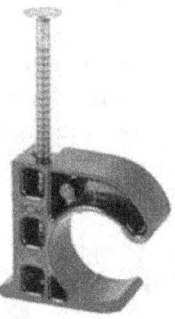

The Sioux Chief Manufacturing Company is one of my personal favorite when it comes to plumbing supports and hangers. The Company was founded in 1957, by Martin E. Ismert Jr. and began in Missouri. In 1953, Martin E. Ismert, Jr., (Ed), left his father-in-law's plumbing supply business and partnered with his brother, Jerry, to form Double I Supply (Double I represented the two Ismert brothers), in Kansas City, Missouri. The company's first proprietary product was the foldable magnesium rod, designed to replace conventional rods for water heaters. Ed Ismert and his brother, Ted Ismert, saw a new need in the post- World War II home building market for a newly designed air chamber, which was a required device on water lines to prevent "water hammer" (banging when valves are turned off quickly). Most municipalities required the air chambers, although the length and diameter could vary from place to place. Plumbers would fabricate the air chambers on site, and the labor-intensive installation processes that would include a copper cap, solder, torch, and flux.

Vertical piping must be supported as to keep the piping in alignment. The support should be strong enough to carry the weight of the full capacity of the pipe being supported, or in the case, the pipe is completely full of liquid. Stacks must be supported at the base and in stories of two or more need to be supported at every floor.

Cast iron: shall be supported in suspended pipe at no less than 18 inches from each hub, and no more than five-foot intervals. Pipe exceeding five feet must be supported at not more than ten feet. Hubless or gasket pipe must be supported at least every other joint, except when the distance between the hub is more than four feet. Supports shall be at every joint.

Threaded pipe: pipe size of 1-1/2 inches and larger shall be supported at 12-foot intervals. Smaller pipe shall be supported at eight-foot intervals.

Copper pipe: hard drawn copper tube shall be at least every eight feet for one inch and smaller tubing. At least ten foot intervals for large tube. Annealed copper tube shall be supported at every eight feet.

Plastic pipe: hangers and straps shall not compress, distort, cut or abrade the piping and shall allow free movement of the piping. Wire hooks shall not be used in supporting plastic pipe. Restraining joints and expansion joints shall be installed as required. (Per manufactures instructions)

All horizontal pipe shall be supported at intervals of not more than four feet, at ends of branches, and change of direction or elevation. Trap arms in excess of three feet shall be supported as close to the trap as possible.

10 WATER SUPPLY AND DISTRIBUTION

Quality of Water

Water quality, unbelievably, is a statement not easily defined. Like many things in nature, water changes as the seasons change. Generally, water quality is referred to in a technical manner, meaning the overall 'scientific' quality of the water. However, water quality can also mean something else. For scientific and legal purposes, the following definition is most often used: Water quality is the ability of a water body to support all appropriate beneficial uses.

Water quality can often be defined in terms of the chemical, physical, and biological content of water. The water quality of rivers and lakes changes with the seasons and geographic areas, even when there is no pollution present. Oddly enough, there is no single measure that constitutes good water quality. For instance, water suitable for drinking can be used for irrigation, but water used for irrigation may not meet drinking water guidelines.

Human health ambient water quality criteria represent specific levels of chemicals or conditions in a water body that are not expected to cause adverse effects to human health. EPA provides recommendations for "water + organism" and "organism only" human health criteria for states and authorized tribes to consider when adopting criteria into their water quality standards. These human health criteria are developed by EPA under Section 304(a) of the Clean Water Act.

EPA sets legal limits on over 90 contaminants in drinking water. The legal limit for a contaminant reflects the level that protects human health and those water systems can achieve using the best available technology. EPA rules also set water-testing schedules and methods that water systems must follow.

The Safe Drinking Water Act (SDWA) gives individual states the opportunity to set and enforce their own drinking water standards if the standards are at a minimum as stringent as EPA's national standards.

[EPA press release – March 8, 1973]

Environmental Protection Agency Deputy Administrator Robert W. Fri today said that the Administration's proposed Safe Drinking Water Act would provide an effective solution to the problem of providing safe drinking water to the public.
Testifying before the House Subcommittee on Public Health and the Environment of the Committee on Interstate and Foreign Commerce, Fri said, "deficiencies among the Nation's drinking water supply systems and surveillance programs have been well documented..." Under the President's bill, Fri testified, the EPA Administrator would establish new Federal primary drinking water standards protective of public health and secondary standards for such matters as taste, odor, and appearance.
State and local agencies would be responsible for determining the necessary controls to meet the standards.
While treatment, operation, maintenance, and construction requirements are not made mandatory under the bill, information as to such would have to be published when standards are issued and would therefore be available for use by suppliers of drinking water in their operation. "That information could also be readily converted into regulations by State and local governments," Fri said. "We believe the enforcement provisions of the bill will be highly effective, almost self-executing and require little direct Federal involvement, he said.
The bill provides that the States shall have primary enforcement authority with regard to the drinking water standards and that EPA will monitor activities of the States and public water supply systems only to the extent necessary to determine if there is an adequate program to enforce the primary standards. Whenever the water delivered by a water supply system does not comply with the primary drinking water standards, the supplier of the water must notify its users, appropriate State agencies, and EPA of the non-compliance and the possible health effects of such non- compliance.

[EPA Journal – March 1975]

The Safe Drinking Water Act of 1974 put into motion a new national program to reclaim and ensure the purity of the water we consume. Under the Act, each level of government, every local water system, and the individual consumer have well-defined roles and responsibilities.
However, both the opportunity and the challenge of implementing the Act begin with each of us in EPA.
The urgency of the task is underscored not only by stringent deadlines in the Act by recent questions about the health effects of chemicals in drinking water. President Ford Signed the Safe Drinking Water Act December 16, 1974, in the wake of newspaper headlines, television documentaries, and magazine features warning that our old assumptions about the quality of our drinking water may no longer be valid.

Potential cancer-causing chemicals have been found in trace quantities in New Orleans and Pittsburgh's drinking water. In Boston, lead from water supply pipes has been found in water drawn from the tap. Viral or bacteriological contamination of drinking water has resulted in communication of disease, often in smaller, more rural communities where treatment works are outdated or modern techniques are not available.

In no case, however, can the interim primary regulations go unenforced beyond December 1976, when they become law for every public water supply system regardless of whether a State has assumed enforcement responsibility.

In cases where a State fails to assume this authority, or fails to exercise it adequately, the Administrator may, after notice to the State, seek a mandatory compliance with these standards through the courts. In any case, the non-complying system must give public notice of its non- compliance to each of its users and to the news media.

The consumer becomes an enforcer and can exert pressure on the utility, the local government, and the State, demanding water that complies with the Federal and State regulations. The Safe Drinking Water Act has real "teeth" from the Federal level down to each of us as consumers.

[EPA press release – June 25, 1977]

National safe drinking water standards go into effect today across the country. Environmental Protection Agency regulations require that the Nation's 40,000 community drinking water systems and 200,000 other public water systems test their water on a routine basis to make sure it is safe to drink. A novel aspect of the law requires utilities to notify consumers if the health standards or sampling requirements are not being met.

"Today marks an important milestone in our effort to insure the safety of the Nation's drinking water supplies," said EPA Deputy Administrator Barbara Blum. "Beginning today, water systems not already doing so will have to initiate programs to sample their water on a routine basis to make sure it meets the interim standards. Many systems already meet these requirements because of existing State programs.

"The 'public notification' provision of the regulations is the most novel feature of the new drinking water program," Blum said. "In the future, consumers will know when their water supply systems encounter problems, and they can help determine solutions including how to pay for the improvements." The regulations going into effect today set health standards for microbiological contaminants, ten inorganic chemicals, six organic pesticides, turbidity (or murkiness) and radiological contamination. These are the first health-related drinking water standards to apply to virtually all public water systems coast to coast.

To assist States in moving toward the assumption of primacy, EPA has dispersed $25 million to help set up and maintain adequate State programs.

An additional $20.5 million will be available for Fiscal Year 1978, which begins on October 1 of this year.

Approximately 35 States have enacted new or revised drinking water legislation and 44 States have modified, or are in the process of modifying, their drinking water regulations. State expenditures for drinking water activities have increased by about $4.8 million (32 percent) since passage of the Safe Drinking Water Act in 1974.

[EPA press release – June 20, 1986]

President Ronald Reagan yesterday signed the 1986 Amendments to the Safe Drinking Water Act, the first major environmental legislation to become law this year. Lee M. Thomas, Environmental Protection Agency Administrator, said, "This law greatly increases EPA's responsibilities for protecting the nation's drinking water. We intend to faithfully carry out its provisions to assure the continued safeguarding of this precious resource."

The measure also provides substantial new authority to EPA to enforce the law including increased civil and criminal penalties for violations.

The new law:

- Requires EPA to regulate more than 80 contaminants in drinking water within three years, and, after that, at least 25 more by 1991.
- Requires certain water systems using surface water sources to use filtration treatment under appropriate circumstance.
- Requires certain water systems using groundwater sources to use disinfection treatment.
- Calls for EPA to impose new monitoring requirements on public water systems for contaminants not yet regulated.
- Provides for a demonstration program to protect critical portions of designated aquifers.
- Requires states to develop programs for protecting areas around wells supplying public drinking water systems.
- Requires EPA to issue new rules for monitoring wells injecting wastes below drinking water sources, and report to Congress on other types of injection wells.
- Prohibits use of lead solders, flux and pipes in public water systems. States will enforce this provision.

August 6, 1996

The President today will sign into law new legislation to ensure American families have clean, safe tap water. In his remarks, he will emphasize the fundamental importance of providing Americans with confidence that they have safe food and water. He will also highlight the bipartisan support for this bill, which allows American families that security. The Safe Drinking Water Act Amendments of 1996 strengthen and expand the nation's drinking water protections. The new law is based on the Administration's 1993 reform proposal including a drinking water treatment loan fund and protection for drinking water sources.

The Clinton Administration proposed improvements in consumer information about local tap water in 1995. The Administration actively supported strong requirements now in the new law that will make more information public than ever before, giving Americans access to direct, simple information–sent directly to their homes in water utility bills–about local water quality, contaminants, water sources, and whether the water poses a risk to health. The new law will strengthen standards that will protect Americans from contaminants that pose the greatest health risks–a principle already embraced in the President's budget. The law sets clear schedules for developing standards for deadly microbial contaminants like cryptosporidium, and calls for considering special populations such as the elderly, children, people with HIV/AIDS, and others when standards are set, to ensure stronger health protection. The law also mandates technical assistance to help water systems nationwide do a better job of delivering safe, clean water.

Providing safe drinking water is a partnership that involves EPA, the states, tribes, water systems, and water system operators. The public drinking water systems regulated by EPA and delegated states and tribes provide drinking water to 90 percent of Americans.

A public water system provides water for human consumption through pipes or other constructed conveyances to at least 15 service connections or serves an average of at least 25 people for at least 60 days a year. A public water system may be publicly or privately owned. There are approximately 155,000 public water systems in the United States. EPA classifies these water systems according to the number of people they serve, the source of their water, and whether they serve the same customers year-round or on an occasional basis.

Colorado River

The city of Durango and La Plata County, Colorado, have declared a state of emergency after a federal cleanup crew accidentally released mine waste into the water.

An estimated 1 million gallons of waste water spilled out of an abandoned mine area in the southern part of the state, turning the Animas River orange and prompting the Environmental Protection Agency to tell locals to avoid it.

"This action has been taken due to the serious nature of the incident and to convey the grave concerns that local elected officials have. To ensure that all appropriate levels of state and federal resources are brought to bear to assist our community not only in actively managing this tragic incident but also to recover from it," said La Plata County Manager Joe Kerby.

According to the EPA, the spill occurred when one of its teams was using heavy equipment to enter the Gold King Mine, a suspended mine near Durango. Instead of entering the mine and beginning the process of pumping and treating the contaminated water inside as planned, the team accidentally caused it to flow into the nearby Animas River.

Before the spill, water carrying "metals pollution" was flowing into a holding area outside the mine.

Colorado Parks and Wildlife officials have been watching for any effects on wildlife since the incident began on Wednesday. They are optimistic that the effects of the spill on terrestrial wildlife will be minimal, the EPA said.

Fish are more sensitive to changes in water.

Officials said they believe the spill carried heavy metals, mainly iron, zinc, and copper, from the mine into a creek that feeds into the Animas River.

From there, the orange water plugged steadily along through the small stretch of winding river in southern Colorado and across the state border to New Mexico where the Animas meets the San Juan River. The EPA and the New Mexico Environment Department said they would test private domestic wells near the Animals to identify metals of concern from the spill.

The state environment department conducts tests on public drinking water systems separately, the agencies said.

Protecting the Potable Water

2014 Illinois Plumbing Code section (I) 890.1110

All premises intended for human habitation or occupied shall be provided with a potable water supply. The potable water shall not be connected to non-potable water and shall be protected from backflow and back siphonage.

What is backflow?
Backflow means the reversal of water flow from its normal or intended direction of flow. Whenever a water utility connects a customer to the utility's distribution system, the intention is for the water to flow from the distribution system to the customer.

However, the flow of water could be reversed from the customer back into the distribution system. If cross-connections exist within the customer's plumbing system when backflow occurs, then it is possible to contaminate the public water supply. There are two types of backflow – backpressure backflow and back siphonage.

What is backpressure backflow?
Backpressure backflow occurs when the pressure of the non-potable system exceeds the positive pressure in the water distribution lines; that is, the water pressure within an establishment's plumbing system exceeds that of the water distribution system. For example, there is a potable water connection to a hot water boiler system that is not protected by an approved backflow preventer. If pressure in the boiler system increases to a point that it exceeds the pressure in the water distribution system, a backflow from the boiler to the public water system may occur. A downstream pressure that is greater than the potable water supply pressure causes backpressure backflow. Backpressure can result from an increase in downstream pressure, a reduction in the potable water supply pressure or a combination of both.

Boiler pumps, pressure pumps, or temperature increases in boilers can create increases in downstream pressure. Reductions in potable water supply pressure occur whenever the amount of water being used exceeds the amount of water being supplied, such as during water line flushing, firefighting or breaks in water mains.

What is back siphonage?
Back siphonage occurs when there is a partial vacuum (negative pressure) in a water supply system, which draws the water from a contaminated source into a potable water supply. The water pressure within the distribution system falls below that of the plumbing system it is supplying.

The effect is similar to siphoning or drinking water through a straw. For example, during a large fire, a pump is connected to a hydrant. High flows pumped out of the distribution system can result in significantly reduced water pressure around the withdrawal point. A partial vacuum has been created in the system, causing suction of contaminated water into the potable water system. During such conditions, it is possible for water to be withdrawn from non-potable sources located near the fire- for example, air-conditioning systems, water tanks, boilers, fertilizer tanks, and washing machines- into buildings located near a fire. The same conditions can be caused by a water main break.

Garden hoses, toilets, or similar devices create most household cross-connections. Under certain conditions, the flow in household water lines can reverse and siphon contaminates into the water supply. A toilet installed incorrectly without a "plumbing-code approved" toilet ballcock (air gap) will allow contaminated water to backflow to other water outlets in your house, including the kitchen sink.

What is a Backflow Preventer?

A backflow preventer is a method or mechanical device to prevent backflow. The basic method of preventing backflow is an air gap, which either eliminates a cross-connection or provides a barrier to backflow.

Mechanical backflow preventers are devices that provide a physical barrier to backflow. There are four devices commonly used – the reduced pressure principle assembly, the double check valve assembly, the pressure vacuum breaker and the atmospheric vacuum breaker. All of these devices require periodic maintenance and testing.

What is an Air Gap?

An air gap is a vertical, physical separation between the end of a water supply outlet and the flood-level rim of a receiving vessel. This separation must be at least twice the diameter of the water supply outlet and never less than one inch. An air gap is considered the maximum protection available against backpressure backflow or back siphonage, but is not always practical and can easily be bypassed.

What is a Reduced Pressure Principle Assembly (RP or RPBA)?

An RP is a mechanical backflow preventer that consists of two independently acting, spring-loaded check valves. It is a hydraulically operating, mechanically independent, spring-loaded pressure differential relief valve between the check valves and below the first check valve. It includes shutoff valves at each end of the assembly and is equipped with test cocks. An RP is effective against backpressure backflow and back siphonage and may be used to isolate health or no health hazards. The RP may be used on all direct connections that may be subject to backpressure or back siphonage and where there is the possibility of contamination by the material that does constitute a potential health hazard. A health hazard or high hazard is a cross-connection involving any substance that could cause death, illness, spread disease, or have a high probability of causing such effects. The degree of hazard refers to a contaminant being toxic on nontoxic, whereby a health hazard involves a toxic substance.

Are a Double Check and Dual Check Backflow Preventer the same thing?
NO. They sound very similar, and they are "relatives," but they are not the same. A Double Check will ALWAYS have two manual shutoff valves – one on the inlet and one on the outlet. These manual valves are used as emergency shut-offs and are necessary to properly test the operation of the backflow preventer. A Double Check will also have test cocks (small outlets) for connecting the test gauges. If it does not have those shut off valves and test cocks, it is NOT a Double Check Backflow Preventer.
A dual check is more of a flow-control device rather than a backflow preventer. Dual check devices do not have shut-off valves or test cocks.

What is a Pressure Type Vacuum Breaker (PVB)?
A PVB is an assembly consisting of an independently operating, internally loaded check valve and an independently operating, loaded air-inlet valve located on the discharge side of the check valve. The device includes tightly closing shut-off valves on each side of the check valves and properly located test cocks for the testing of the check valve(s). PVBs may be used as protection for connections to all types of non- potable systems where the vacuum breakers are not subject to backpressure. These units may be used under continuous supply pressure. They must be installed above the usage point.

What is an Atmospheric Type Vacuum Breaker (AVB)?
The purpose of the AVB is to prevent a siphon from allowing a contaminant or pollutant into the potable water system. They do not prevent backflow from backpressure.
The most commonly used atmospheric type anti-siphon vacuum breakers incorporate an atmospheric vent in combination with a check valve. Its operation depends on a supply of potable water to seal off the atmospheric vent, admitting the water to downstream equipment. If a negative pressure develops in the supply line, the loss of pressure permits the check valve to drop; sealing the orifice, while at the same time the vent opens, admitting air to the system to break the vacuum. AVBs can be used on most inlet type water connections that are not subject to backpressure, such as low inlet feeds to receptacles containing toxic and nontoxic substances, valve outlet, or fixture with hose attachments, lawn-sprinkler systems, and commercial dishwashers.

Below are diagrams of each device.

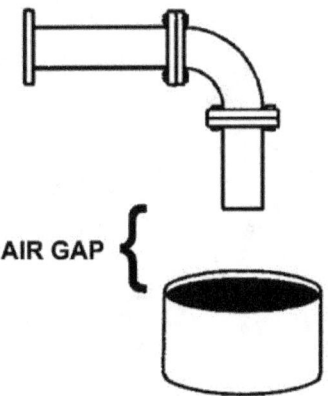

AIR GAP

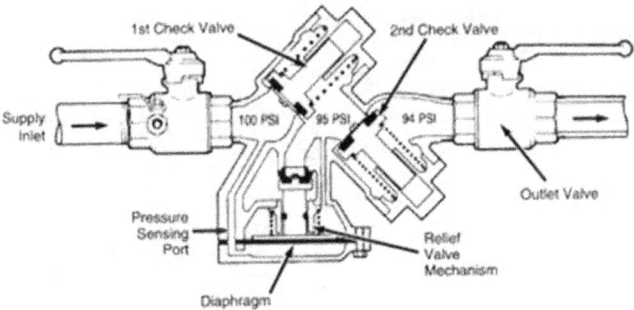

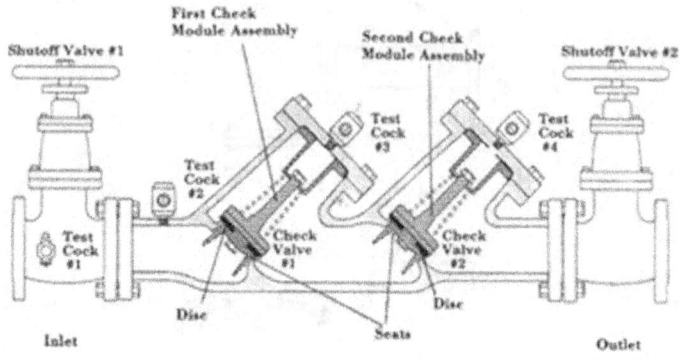

Dual Check Valve

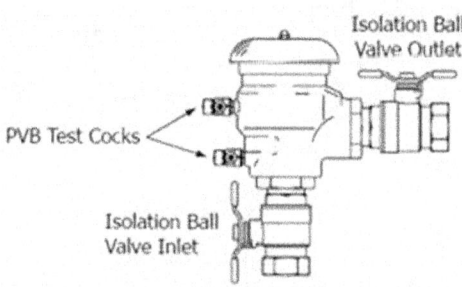

Pressure Vacuum Breaker

Atmospheric Vacuum Breaker

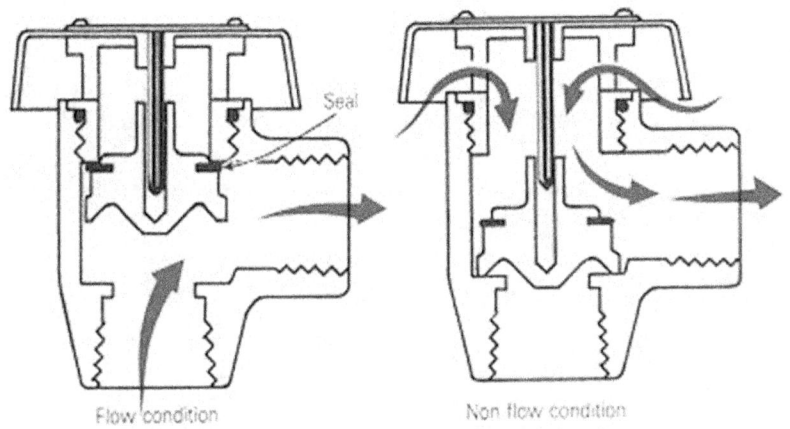

Flow condition — Non flow condition

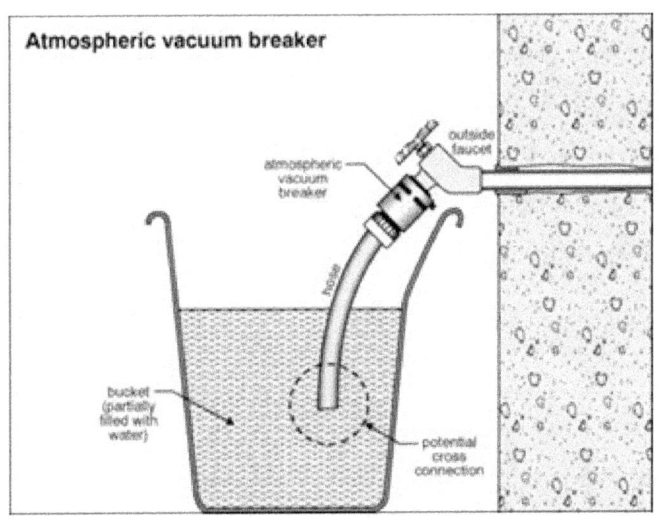

It is important to point out the state of Illinois only recognizes two types of hazards when referring to level of backflow protection. High hazard and low hazard.

Section 890.1130 protection of potable water

Devices for the Protection of the Potable Water Supply. Approved backflow preventers or vacuum breakers shall be installed with all plumbing fixtures and equipment that may have a submerged potable water supply outlet and that are not protected by a minimum fixed air gap. Connection to the potable water supply system for the following fixtures or equipment shall be protected against backflow with one of the appropriate devices as indicated below:

1) Inlet to receptacles containing *low hazard* substances (steam, compressed air, food, beverages, etc.):

A) fixed air gap fitting;

B) reduced pressure principle backflow preventer assembly;

C) atmospheric vacuum breaker unit;

D) double check valve backflow preventer assembly;

E) double check backflow preventer with atmospheric vent assembly; or

F) dual check valve.

2) Inlet to receptacles containing *high hazard* substances (vats, storage containers, plumbing fixtures, etc.):

A) fixed air gap fitting;

B) reduced pressure principle backflow preventer assembly; or

C) atmospheric vacuum breaker unit.

3) Coils or jackets used as heat exchangers in compressors, degreasers and other equipment involving *high hazard* substances:

A) fixed air gap fitting; or

B) reduced pressure principle backflow preventer assembly.

4) Direct connections that are subject to back pressure:

A) Receptacles containing low hazard substances (vats, storage containers, plumbing fixtures, etc.):

i) fixed air gap fitting;

ii) reduced pressure principle backflow preventer assembly;

iii) double check valve backflow preventer assembly;

double check backflow preventer with atmospheric vent assembly; or

iv) dual check valve.

B) Receptacles containing high hazard substances (vats, storage containers, etc.):

i) Fixed air gap fitting; or

ii) A reduced pressure principle backflow preventer assembly.

5) Inlet to or direct connection with sewage or lethal substances: fixed air gap fitting.

6) Hose and spray units or stations shall be protected by one of the appropriate devices as indicated below:

A) Fixed air gap;

B) Reduced pressure principle backflow preventer assembly;

C) Double check valve backflow preventer assembly;

D) Double check valve backflow preventer with atmospheric vent assembly;

E) Dual check valve backflow preventer assembly;

F) Atmospheric vacuum breaker unit.

g) Installation of Devices or Assemblies

1) Devices of All Types. Backflow preventer assemblies and devices shall be installed to be accessible for observation, maintenance, and replacement services. Backflow preventer devices or assemblies shall not be installed where they would be subject to freezing conditions, except as allowed in Section 890.1140(d).

2) All in-line backflow/back siphonage preventer assemblies shall have a full port type valve with a resilient seated shut-off valve on each side of the preventer. Relocation of the valves is not permitted.

3) A protective strainer shall be located upstream of the first check valve on all backflow/back siphonage preventers unless the device contains a built-in strainer. Fire safety systems are exempt from the strainer requirement.

4) Atmospheric vacuum breakers shall be installed with the critical level above the flood level rim of the fixture they serve, and on the discharge side of the last control valve of the fixture. No shut-off valve or faucet shall be installed beyond the vacuum breaker.

5) No in-line double check valve backflow preventer assembly (DCV) or reduced pressure principle backflow preventer assembly (RPZ) shall be located more than 5 feet above a floor, or be installed where it is subject to freezing or flooding conditions. After installation, each DCV and RPZ shall be field-tested in-line in accordance with the manufacturer's instructions by a cross-connection control device inspector before initial operation. (See subsection (b).)

6) A dual check backflow preventer with atmospheric vent assembly shall not be installed where it is subject to freezing or flooding conditions.

7) Closed water systems with hot water storage shall have a properly sized thermal expansion tank located in the cold water supply as near to the water heater as possible and with no shut-off valve or other device between the heater and the expansion tank. Exception: In existing buildings with a closed water system, a properly sized pressure relief valve may be substituted in place of a thermal expansion tank. For closed water systems created by backflow protection in manufactured housing, as required in Section 890.1140(i), a ballcock with a relief valve may be substituted for the thermal expansion tank.

(Source: Amended at 38 Ill. Reg. 9940, effective April 24, 2014)

Backflow/What, Could Go Wrong?

DATE OF BACKFLOW INCIDENT: *June 1991* LOCATION OF BACKFLOW INCIDENT: *Casa, Arkansas* SOURCE(S) OF

INFORMATION: - *Pacific Northwest Section of the American Water Works Association, Summary of Backflow Incidents, Fourth Edition, 1995 – Watts Industries, Inc.; Watts Regulator News/Stop Backflow*

During the week of June 23, 1991, residents near a poultry farm in Casa, Arkansas, became concerned when their water appeared discoloured. In response to complaints from one water customer, it was discovered that the public water system had been contaminated by backflow from a chicken house at the poultry farm. Both the public water system and auxiliary water well supplied water to the plumbing in the chicken house. The water service connection from the public water system to the chicken house included two single check valves in series for backflow prevention. Workers were using the water in the chicken house to administer an antibiotic solution to the chickens.

When the Casa water system manager became aware of the problem, the manager shut off water service to the chicken house and flushed the public water main serving the area. He later removed the water meter serving the chicken house until a proper backflow preventer could be installed. The feeding of antibiotic solutions and live virus vaccines into water to treat and immunize chickens is a popular practice at poultry farms.

Such antibiotic solutions could cause severe adverse effects in humans who are hypersensitive to the drugs, and most of the virus vaccines used to immunize chickens is pathogenic to humans. Therefore, poultry farms should be considered a significant health hazard to public water systems, and a reduced-pressure principle backflow-prevention assembly should be installed at the water service connection to each poultry farm.

Backflow at a Mortuary
SOURCE(S) OF INFORMATION: - U.S. Environmental Protection Agency, Cross-Connection Control Manual, 1989 CASE HISTORY

The chief plumbing inspector in a large southern city received a telephone call advising that blood was coming from drinking fountains at a mortuary (i.e., a funeral home). Plumbing and health inspectors went to the scene and found evidence that blood had been circulating in the potable water system within the funeral home. They immediately ordered the funeral home cut off from the public water system at the meter. City water and plumbing officials did not think that the water contamination problem had spread beyond the funeral home, but they sent inspectors into the neighborhood to check for possible contamination. Investigation revealed that blood had back flowed through a hydraulic aspirator into the potable water system at the funeral home. The funeral home had been using a hydraulic aspirator to drain fluids from bodies as part of the embalming process. The aspirator was directly connected to a faucet at a sink in the embalming room. Water flow through the aspirator created suction used to draw body fluids through a needle and hose attached to the aspirator. When funeral home personnel used the aspirator during a period of low water pressure, the potable water system at the funeral home became contaminated. Instead of body fluids flowing into the waste water system, they were drawn in the opposite direction–into the potable water system.

Propane in the Water Supply
SOURCE(S) OF INFORMATION: - *Pacific Northwest Section of the American Water Works Association, Summary of Backflow Incidents, Fourth Edition, 1995 – U.S. Environmental Protection Agency, Cross- Connection Control Manual, 1989 – Watts Industries, Inc.; Watts Regulator News/Stop Backflow*

In August 1982, residents in a Connecticut town reported hissing, bubbling noises coming from washing machines, sinks, and toilets. Faucets sputtered out small streams of water mixed with gas. Propane gas had back flowed into the town's public water system. Local firefighters and other officials asked hundreds of residents to evacuate their homes and businesses. The town provided water to a propane storage facility in the area. Water was furnished to the facility for both domestic use and fire protection and entered the facility through a single eight-inch-diameter service connection. The facility included 26 subsurface 30,000-gallon liquid propane storage tanks.

On the day of the backflow incident, workers needed to repair a storage tank at the propane storage facility. Before repairing the tank, workers had to purge the tank of residual propane. There are two common methods for purging liquid propane storage tanks. One method is to use an inert gas such as carbon dioxide. The other method is to use water.

The use of water is the preferred method because it is a more positive method and will float out any sludge as well as gas vapors. Accordingly, workers attempted to purge the tank using water in this case. They connected a hose to the tank from one of the two fire hydrants at the facility. Unfortunately, the pressure in the propane tank was about 85 to 90 psig, while the pressure in the town's public water system was about 65 to 70 psig. Consequently, propane gas back flowed into the town's public water system. It was estimated that about 2,000 cubic feet of gas flowed into the water system over a period of about 20 minutes. This is enough gas to fill approximately one mile of eight-inch diameter water main.

Fires were reported at two houses, and fire gutted one of these houses. At another house, a washing machine exploded. Police, propane company workers, and town water works personnel, however, limited damage and injuries by quickly sealing off the affected area. The town flushed fire hydrants and individual building plumbing systems and monitored for gas. The propane company promptly instituted revised propane tank purging procedures at its storage facility.

Nicor- Aurora, Illinois
Source; The Beacon News

Ruben Luna's doctor was stumped. The Aurora resident's teeth were falling out, he had no control over his bowels and the sharp jabs in his gut left him hunched over daily. —You can't be this sick, the doctor told Luna. —You have the body of an 80-year-old and you're 30.

The doctor gave him pills. Nothing changed. They took out part of his stomach, removed some gallstones. However, the grinding ache persisted. At work, he was known as —The Bone Chewer, since he went after overtime hours like a dog fighting for the last bone on the block. With that reputation, Luna didn't want to tell people he was suffering. So it was a long time before he found out he wasn't alone. At the Nicor facility in Aurora where Luna worked for 12 years, plenty of others were in pain. According to a lawsuit filed in 2004, dozens of people who worked at the Nicor building at 408 S. River St. developed liver failures, suffered persistent diarrhea and threw up every day. Tom Schultz, a part-time martial arts instructor with two kids, had three urinary tract infections and doctors used sound waves to bombard kidney stones out of him. At night, the 40-year employee's agony made him think about suicide.

Health nut Anna Sutton, 44, was plagued by migraine headaches and struggled with bladder control. Libby Thompson, 42, found groups of small lumps under the skin of her neck, stomach, and pelvis. Health nut Anna Sutton, 44, was plagued by migraine headaches and struggled with bladder control. Libby Thompson, 42, found groups of small lumps under the skin of her neck, stomach, and pelvis.

They all believe faulty plumbing that city records show could have allowed chemicals to leak into the drinking water caused their suffering. It's a problem health officials believe could be responsible for hundreds of illnesses across the country every year, sickening people who drink from the water fountains in their offices, or using contaminated water to wash their food.

Bruce Brummel, a former Nicor employee, has been working for years to get the company to acknowledge what city of Aurora documents already show: Bad plumbing in the building's break room meant employees could have been sipping water tainted with three times the legal levels of methylene chloride.

According to the Environmental Protection Agency, methylene chloride can do damage to the nervous and blood systems during short exposures. Over years, experts believe it causes liver damage, cancer and destroys the digestive system. For the workers in the street — the men and women installing and repairing the gas lines for Nicor — the job was a great opportunity. Although most of them had no education past high school, dedicated employees could make a sizable salary.

The street crew became a tight-knit, informal club, and their camaraderie meant it wasn't unusual to find a handful of workers gathered in the company's break room, sipping coffee from foam cups an hour before their shift started. A regular part of the morning chatter was griping about the coffee.

While those jokes are common at most workplaces, the brew in the Nicor break room seemed to exceed bad flavor. It was grimy, thick and smelled a little like oil, employees say. —I think I said it first: I said this water tastes like boiler water, said one current Nicor employee, who asked to remain anonymous. —It was a joke.

However, the comment got people thinking. Brummel says he and a co-worker traced the pipe back to its source. According to city of Aurora records, the pipe was coming straight out of the building's boiler, a massive piece of machinery used to supply heat to the offices.

In a typical plumbing system, water comes in from the city's main pipes, then splits into drinking water and non-potable water used for things like fire prevention sprinklers or, in the case of Nicor, to run the boiler. In those non-potable areas, the water mixes with various chemicals. In a safe system, any place chemicals mix with the water, safety devices called —backflow preventers must be installed to ensure the tainted water can't move back into the drinking supply. Health officials believe that every year hundreds of people across the nation are sickened by drinking water contaminated in plumbing that lacks backflow protection. Unlike the Nicor employees, most victims suffer short-term but severe symptoms, which they attribute to the 24-hour flu or —something they ate. Health experts and plumbers contend holes in the state plumbing code or incomplete testing are just as likely culprits. —It's a little scary to realize how vulnerable the water systems are.

Said Lou Allyn Bius, manager of the Division of Public Water Supplies for the Illinois Environmental Protection Agency, who retired in 2005. —It's so easy to connect something that can quickly overwhelm the whole system. The average person probably isn't aware of it all.

According to city documents, pipes at Nicor that were supposed to be one-way could allow water to mix with toxic chemicals from the boiler during low water pressure. Brummel felt he might have discovered an explanation for the bad coffee — and a few other odd occurrences at the office. The most obvious was the daily lines at the men's rooms. Every morning, shortly after drinking coffee, the workers would rush to the bathroom. Men in their 20s and 30s describe lining up three-deep for a stall or running up two floors to make it to an upstairs bathroom before their diarrhea hit. However, no one talked about it. Brummel began subtly asking around — and quickly found similar stories. Otherwise, healthy young men and woman were pulling over to vomit after leaving the Aurora building. Others had defecated in their pants on the job. Dozens of workers had severe weight loss, debilitating headaches and rotator cuffs that were disintegrating.

Brummel, always thin, dropped 35 pounds suddenly. His eyes sunk into his head, rashes appeared on his chest. At 35, Brummel's joints creaked and crackled like an old man's.

Brummel's doctor said his symptoms appear to be consistent with chemical poisoning. —A person who appears somewhat emaciated and has symptoms affecting virtually every organ in their system, it's often chemically involved, said Pauline Harding, a Winfield physician who has been treating Brummel. Harding said toxins like mold in the garage assault us all, but it takes something extra to do real damage. —Most of us manage to survive those, but if you get an assault by chemicals in your drinking water, it can be overwhelming.

Brummel says he took his concerns to his immediate supervisor. He and coworkers went up the management chain, but at each step, the company denied there was a problem, Brummel contends.

After months of working within the company and still unable to get help, on Oct. 14, 2003, Brummel took pictures of the company's plumbing to Aurora's Division of Building and Permits. According to Brummel and to city records, within 20 minutes, officials from the city of Aurora descended on Nicor's building to investigate Brummel's claims about the drinking water. By all accounts, all sides slung heated accusations that day. It ended with Brummel — who was on sick leave at the time — being banned from Nicor property for the rest of the week. It also resulted in an investigation by the Occupational Safety and Health Administration, a federal organization that promotes safe working environments. An OSHA report started that day shows Nicor was made aware of a potentially dangerous plumbing situation in its building.

According to the OSHA report, on Oct. 14, the city of Aurora took a sample of tap water from the Nicor plant to be tested by an independent agency. That test detected methylene chloride levels at
17.5 parts per billion — more than three times the limit of five parts per billion allowed by the Environmental Protection Agency, the report shows. Methylene chloride is a colorless chemical compound that's typically used to strip paint. In past decades, methylene chloride gas was used in Christmas lights and jukeboxes to create a bubbling effect, until health risks led companies to search for safer alternatives. According to plumber Marty Feltes, who viewed the boiler with city inspectors on Oct. 14, methylene chloride was most likely being used to protect the pipes, slowly being injected to erode build-up from steam production. Questions were raised regarding the test but, according to the OSHA report, a second test on Oct. 24, this time run by Nicor, showed methylene chloride levels at 16.1 parts per billion, still three times the limit allowed by the EPA. Citing the violation as —serious, OSHA was prepared to fine the company $2,125, OSHA records show.

Later, a third test, according to OSHA records, showed methylene chloride at 0.89 parts per billion — below the legal limit of five parts per billion — and the fine was not imposed. Shortly after that Oct. 29 test, OSHA closed its investigation, saying the agency could not prove the case, although the agency did not retract its findings.

A week after Nicor got test results showing methylene chloride above legal limits, the company also received a warning letter from the city. The Oct. 21 letter, written by city of Aurora plumbing inspector Robert Thompson to Nicor's building supervisor, said the company was in violation of the Illinois

State Plumbing Code.

After inspecting the boiler room, Thompson wrote that he found the drinking water was directly connected to a chemically fed tank. —This type of connection poses a threat to the quality of potable water for both the employees in the building and the city's water main, Thompson's letter said. Thompson advised Nicor to immediately have a licensed plumber install backflow protection to stop the drinkable and dangerous water from mixing, or permanently disconnect the pipe, city records show. However, city and company documents show after Nicor officials received the city's letter. They continued to assure employees the water was safe while also trying to fix plumbing problems the city had deemed hazardous.

Less than a month after the city's alert, Nicor posted a memo to employees telling them the water was safe to drink. And, on Nov. 19, 2003, the company's vice president of human resources, in a letter to Brummel, wrote that an investigation by OSHA had determined boiler piping —was not improperly connected to the potable drinking water.

The letter went on to assert that no —violations of any safety or health regulations relating to company drinking water‖ were found. OSHA records appear to indicate that the spigot where workers were routinely filling up their water jugs and coffee pots was not considered drinking water by Nicor, so could not be investigated by OSHA in the same way.

Then, on Nov. 25, Aurora city records show Nicor's area manager applied for a remodeling permit to correct a problem the company denied existed. The application notified the city that, for the cost of
$1,600, Nicor would add a backflow protection device to each service entering in the building. Without backflow protection, the building was in violation of the Illinois State Plumbing code, which has required a safeguard system since 1983. City documents show the backflow devices were finally installed by January of 2004. Yet, for more than three years, even as workers protested outside their headquarters, the company has steadfastly denied any problems with the water. Nicor refused repeated requests for an in-person interview with The Beacon News, but issued two written statements, one through the corporate spokesman and one from the company's legal department.

Martinez, the spokesman, said all Nicor buildings do, and have always, —operated in compliance with all applicable codes.‖ —It is apparent these allegations of a hazard were found not to have merit almost two years ago, said Jill Kelly, assistant legal counsel representing Nicor.
—Accordingly, we trust that any story that is forthcoming will not inaccurately assert that any hazard existed at the facility. The company refused any further comment on the OSHA or city reports that showed serious plumbing issues. After years of good reviews, Brummel was fired in April 2004. He believes it is because he tried to expose what he saw as a dangerous problem at Nicor's building.

He filed grievances through his union (which were dropped, he said) and held candlelight vigils for the sick employees. Chicago law firm Cascino Vaughan

Law Offices Ltd., filed suit on behalf of Brummel and 32 other current or former Nicor employees.
Partner Allen Vaughan said there could be complications with the lawsuit because some of the older workers may have been exposed to mercury while working at Nicor. The attorneys held public meetings, wrote to local politicians. After months of work, the lawyers withdrew from the case, saying the case was not economically feasible to pursue. Since then, Brummel and other Nicor current and former employees have been working with Waukegan attorney Rick Daniels. Daniels said he is still in the investigation phase of his work, but confirmed his firm, Daniels, Long and Pinsel, is prepared to file a federal suit.
Daniels said he is not intimidated by the cost or complexity of a suit. His only worry was that all of the plaintiffs would see the end of their efforts. —It's going to be a huge case, one of the biggest in Illinois history, Daniels predicted. —It's an absolute tragedy what happened.‖ State Rep. Linda Chapa LaVia, an Aurora Democrat whose sister worked for Nicor, has also met with some current and former Nicor employees. —If there aren't laws protecting people from this. I plan to pull together some legislation, she said. —There's just too many coincidences surrounding Brummel and others. People need to know everyone's life matters. In this case, we have quite a few coincidences to quite a few people in a fixed amount of time that you can't overlook. Nevertheless, as those legal and legislative wheels start to turn, workers are still suffering, almost three years after the city inspection.

Schultz, the former karate instructor, is constantly on the toilet. Doctors still can't get rid of the lumps under Thompson's skin. She left for a new job in 2004, but her lungs are so weak it's hard for her to walk upstairs. In addition, Luna, 49, can't work at the landscaping company he started after he left Nicor. —My biggest fear is that I'm sick and my family will have to take care of me like a baby, he said. —I told them just let me die.

Water into Wine
SOURCE(S) OF INFORMATION: - *Pacific Northwest Section of the American Water Works Association, Summary of Backflow Incidents, Fourth Edition, 1995*

At a winery in the City, someone inadvertently left open a water supply valve to a wine-distilling tank after flushing out the tank. During a subsequent fermenting process, wine back flowed from the tank into the City water mains and out of the faucets of nearby homeowners.
This reversal of flow through the water piping occurred because the pressure in the wine distilling tank was greater than the pressure in the City water system.

Antifreeze Taints Water at school Drinking fountains poison eight students
By Jim Kirksey Denver Post Staff Writer

Eight Brighton middle-school students were sent to the hospital yesterday after drinking antifreeze that had seeped into the water in the school's drinking fountains. Overland Trail Middle School was closed after 11 a.m. yesterday, and it will remain closed today while authorities seek the source of the pollution. Rodger Quist, principal of the 6-year- old school, said the ethylene glycol substance is used in the building's hot water heating system, which isn't supposed to be connected to the drinking water system at Overland Trail.

Pest Control Chemicals in the Water
Summary of Backflow Incidents, Fourth Edition, 1995 – Watts Industries, Inc.; Watts Regulator News/Stop Backflow

On June 24, 1987, a construction crew inadvertently broke a water main while widening a bridge in New Jersey. Several hours after the water main was repaired, a customer called the water department to complain that the water was milky and smelled bad. Pesticides had back flowed into the public water system. The backflow incident happened at the time the bridge construction crew broke the water main. Because of the water main break, a siphoning action occurred in the water mains. Concurrently, a pest control company employee was rinsing a tank that contained a weak solution of the pesticides heptachlor and chlordane. The hose that the employee was using had the pesticide Dursban on it. One to three gallons of the pesticides were sucked through the pest control company's potable water system and into the public water system. Several people drank, and watered their gardens with, the contaminated water. Fortunately, however, there were no immediate illnesses or injuries. After receiving the complaint about milky and bad smelling water, the water department immediately shut off the water supply to the 63 customers affected by the water main break and notified them not to drink the water or use it to cook, bathe, or wash clothes.
The 63 homes and businesses went without usable water service for several days while affected water mains and plumbing were flushed and disinfected. A tank truck provided potable water for drinking and cooking. Shower facilities at the local public high school and middle school were made available for use by affected residents. Because the pesticides stuck to piping, the plumbing at nine locations had to be replaced. At all other locations, analysis of water samples showed that the pesticides were not detectable. The pest control company assumed responsibility for the backflow incident and paid for the necessary replacement of plumbing.
Nevertheless, 21 homeowners sued the pest control company for $21,000,000. They claimed that the pest control company irreparably damaged plumbing fixtures, that residents continue to suffer physical injury, and that residents have been subjected to mental distress, Inconvenience, and loss of property. Pay medical expenses incurred because of the incident and to maintain a health surveillance program for affected residents. The water department ordered the pest control company to cease operating until a

backflow preventer was installed at the water service connection to the pest control company. Following installation of a backflow preventer, the pest control company resumed operating.

About the Environmental Resources Training Center

In 1977, the Illinois Environmental Protection Agency (IEPA) designated the Environmental Resources Training Center (ERTC) as the Illinois center for the continuing education of personnel involved in the operation, maintenance, and management of drinking water and wastewater treatment systems. In 1983/84, the Illinois Environmental Protection Agency additionally designated ERTC to train Illinois licensed plumbers and Illinois certified water operators to become IEPA certified Cross-Connection Control Device Inspectors. ERTC courses are designed to assist both entry-level personnel who are preparing for a career in drinking water and wastewater treatment systems and persons already employed in such systems who desire education to upgrade job skills, obtain advanced certification levels, and prepare for positions that are more responsible. In addition, the ERTC offers courses for licensed plumbers in cross connection control or backflow prevention. Persons who complete ERTC courses are awarded continuing education units (CEUs) by the University and receive education credits applicable to official certification as drinking water or wastewater treatment system operators or in cross connection control under requirements administered by the IEPA.

Flushing/Disinfecting Water

Although several methods eliminate disease-causing microorganisms in water, chlorination is the most commonly used. Chlorination is effective against many pathogenic bacteria, but at normal dosage rates, it does not kill all viruses, cysts, or worms. When combined with filtration, chlorination is an excellent way to disinfect drinking water supplies.

State and federal governments require public water supplies to be biologically safe. The U.S. Environmental Protection Agency (EPA) recently proposed expanded regulations to increase the protection provided by public water systems. Water supply operators will be directed to disinfect and, if necessary, filter the water to prevent contamination from *Giardia lamblia*, coliform bacteria, viruses, heterotrophic bacteria, turbidity, and *Legionella*.

Private systems, while not federally regulated, also are vulnerable to biological contamination from sewage, improper well construction, and poor-quality water sources. Since more than 30 million people in the United States rely on private wells for drinking water, maintaining biologically safe water is a major concern.

Testing Water for Biological Quality

The biological quality of drinking water is determined by tests for coliform

group bacteria. These organisms are found in the
Intestinal tract of warm-blooded animals and in the soil. Their presence in water indicates pathogenic contamination, but they are not considered pathogens. The standard for coliform bacteria in drinking water is "less than 1 coliform colony per 100 milliliters of sample" Public water systems are required to test regularly for coliform bacteria. Private system testing is done at the owner's discretion. Drinking water from a private system should be tested for biological quality at least once each year, usually in the spring. Testing is also recommended following repair or improvements in the well. Coliform presence in a water sample does not necessarily mean that the water is hazardous to drink. The test is a screening technique, and a positive result (more than 1 colony per 100 ml water sample) means the water should be retested. The retested sample should be analyzed for fecal coliform organisms. A high positive test result, however, indicates substantial contamination requiring prompt action. Such water should not be consumed until the source of contamination is determined and the water purified.

Chlorine Treatment

Chlorine readily combines with chemicals dissolved in water, microorganisms, small animals, plant material, tastes, odors, and colors. These components "use up" chlorine and comprise the chlorine demand of the treatment system. It is important to add sufficient chlorine to the water to meet the chlorine demand and provide residual disinfection.
The chlorine that does not combine with other components in the water is free (residual) chlorine, and the breakpoint is the point at which free chlorine is available for continuous disinfection. An ideal system supplies free chlorine at a concentration of 0.3-0.5 mg/l. Simple test kits, most commonly the DPD colorimetric test kit (so called because diethyl phenylene diamine produces the color reaction), are available for testing breakpoint and chlorine residual in private systems. The kit must test free chlorine, not total chlorine. If a system does not allow adequate contact time with normal dosages of chlorine, super chlorination followed by dechlorinating (chlorine removal) may be necessary. Super chlorination provides a chlorine residual of 3.0-5.0 mg/l, 10 times the recommended minimum breakpoint chlorine concentration. Retention time for super chlorination is approximately 5 minutes.
Activated carbon filtration removes the high chlorine residual.

Shock chlorination

Is recommended whenever a well is new, repaired, or found to be contaminated. This treatment introduces high levels of chlorine to the water. Unlike super chlorination, shock chlorination is a "one time only" occurrence, and chlorine is depleted as water flows through the system; activated carbon treatment is not required. If bacteriological problems persist following shock chlorination, the system should be evaluated.

Types of Chlorine used in Disinfection

Public water systems use chlorine in the gaseous form, which is considered too dangerous and expensive for home use. Private systems use liquid chlorine (sodium hypochlorite) or dry chlorine (calcium hypochlorite). To avoid hardness deposits on equipment, manufacturers recommend using soft, distilled, or demineralized water when making up chlorine solutions.

Liquid Chlorine	Dry Chlorine
household bleach most common form available chlorine range: 5.25% (domestic laundry bleach) 18% (commercial laundry bleach) slightly more stable than solutions from dry chlorine protect from sun, air, and heat	powder dissolved in water available chlorine: 4% produces heavy sediment that clogs equipment; filtration required dry powder stable when stored properly dry powder fire hazard near flammable materials solution maintains strength for 1 week protect from sun and heat

Equipment for Continuous Chlorination

Continuous chlorination of a private water supply can be done by various methods. The injection device should operate only when water is being pumped, and the water pump should shut off if the chlorinator fails or if the chlorine supply is depleted.

Design & Sizing Water

The design and installation of hot and cold water within a buildings distribution system must provide a volume of water at the required rates and pressure to ensure the safe, efficient, and satisfactory operation of plumbing fixtures.

A few installation requirements;

A full port shut off valve shall be located near the curb, or property line. A full port valve will also be required at both side of the water meter. A means to drain the system at the meter must also be provided. (Do not forget if you install a boiler drain in which a hose can be connected, you will need to install a vacuum breaker or remove the threads.) Water supply and distribution must also be protected from freezing. The requirement, regarding the depth of a service line as listed in the Illinois plumbing code. At least 36 inches or the maximum frost penetration of the local area. Word of caution, we live in a cold weather climate; I have witness frost as deep as six feet.
Potable water shall have a ten-foot separation (in the horizontal) between sanitary sewer and building sewer. However, the water service and building drain or sewer may be installed in the same trench if the water service is "shelved" a minimum of 18 inches above the building drain, or sewer. If the potable water service passes beneath the sanitary sewer or building drain, the sanitary sewer shall be constructed of one of the materials in Appendix A, table (A) approved building drain. If you flip back to chapter four of this book, I have listed it for you here. The only thing that I will point out is the joint or pipe connections are required to be that of solvent cement. If plastic materials are to be used. This means if you are installing an SDR plastic material for use in the building sewer and you are above the water service you would be required to use schedule 40 PVC with solvent cement joints.
With that in mind you will also be required to extend on each side on each side of the crossing to a distance of at least ten feet as measured at right angles to the water line.
Now for instances where this is not possible, there are alternatives. For an example, you may consider encasing the water service. The casing material

will need to be sealed on both ends and extend ten feet on either side of the center of the sanitary sewer pipe. In addition, the casing or sleeve will need to be two times larger than the water service.

When neither is an option, our codebook reminds us to contact the department for an alternative solution. I have provide contact information at the back of this book. Potable water supply pressure, as it enters a building, is either insufficient or exceeds 80 PSI (pounds per square inch); there will be a need to control pressure by one of two ways. First with inadequate pressure, a booster or auxiliary pressure pump maybe the solution.

However, any installation of a booster pump will require you to process the necessary paperwork to the department for approval.

In the case of supply pressures exceeding 80 PSI, you will be required to install a PRV (pressure reducing valve) at the incoming service line. These devices can be manually adjusted to the desired water pressure.

When calculating the size of a supply or service line, the calculation for demand shall be based on the total number and type of fixtures installed assuming simultaneous use of such fixtures.

When calculating the distribution piping, utilizing the water fixture supply units chart or (WFSU) as with drainage fixture units (chapter 12) Each fixture in plumbing has a specific demand, draw, or load onto the system.

Whenever you are determining the size of a service line and more specifically the piping from the curb-stop, or "B" box. Section 890.1200 explains the use of Appendix A, tables M, N, O, and Q should be used. In addition, it states that the minimum for a service line shall be not less than ¾-inch pipe.

Agency note: Appendix (A) table, (M): For fixtures not listed, loads shall be assumed by comparing the fixtures to one listed using water similar quantities at similar rates.

The assigned loads for fixtures with both hot and cold water supplies are given for separate cold and hot water loads and for total load.

Where local government or the community public water supply does not require separate water service lines for irrigation or similar systems that are likely to impose continuous demands (e.g., lawn sprinkler or air conditioning systems), the following rule applies.

Estimate the continuous demand (in gallons per minute) for outlets/systems separately from the intermittent demand from the fixtures listed in this table, and add this amount to the demand of the fixtures (in gallons per minute). Fire sprinklers are exempt from this table. (Source: Amended at 38 Ill. Reg. 9940, effective April 24, 2014)

* (1 WSFU = 1 GPM) (Uniform Plumbing Code)

> The total theoretical demand for a water supply system can be calculated by adding known maximum demand for all fixtures in the system. Due to the nature of intermittent use, this will unfortunate add up to unrealistic demands for the main supply service lines. A realistic demand for a supply system will always be far less than the total theoretical demand.

Expected demand in a water supply system can be estimated as
$$Q_{et} = q_{nl} + 0.015 \, (\Sigma q_n - q_{nl}) + 0.17 \, (\Sigma q_n - q_{nl}) \, 1/2 \qquad (1) \text{ where}$$
Q_{et} = expected total water flow (l/s) q_{nl} = demand of largest consumer (l/s)
Σq_n = total theoretical water flow – all fixtures summarized (l/s)

Note that minimum expected total water flow can never be less than the demand from the largest fixture. This equation is valid for ordinary systems with consumption patterns such as,
- homes
- offices
- nursing homes
- Etc.

Be aware when using the equation for systems serving large groups of people where the use is intermittent such as,
- hotels
- hospitals
- schools
- theaters
- wardrobes in factories
- Etc.

For these kind of applications, like a wardrobe, it is likely that all showers are used at the same time. Using the formula blindly would result in insufficient supply lines.

Agency Note: Appendix A table, (N) 2014 Illinois Plumbing Codebook

Meter and meter yoke sizes shown in this table shall apply only to those jurisdictions or governmental units where local ordinances or community public water supply requirements do not prescribe specific sizes of meters or meter yokes. Where local ordinances or community public water supply requirements cover sizing, those requirements shall be followed. (Source: Amended at 38 Ill. Reg. 9940, effective April 24, 2014)

A few additional notes regarding water distribution installation considerations. The minimum water at the discharge side or house side of the meter shall be at least 20 PSI. With a minimum constant pressure at each fixture of at least 8 PSI.

When controlling water hammer, all building water supply systems shall be provided with air chambers or approved mechanical device or water Hammer arrestors to absorb pressure surges. Absorbers shall be at the ends of long runs or near batteries of fixtures. An air chamber installed at a fixture shall be at least 12 inches in length and be of the same size as the fixture supply, or an air chamber with equivalent volume may be used. An air chamber that is installed in a riser shall be at least 24 inches in length and of the same size as the riser.

It is also important to note that a dead end must be avoided in the water distribution piping as to prevent stagnant water. A dead end is any pipe longer than 24 inches in length.

Over the next few pages, for quick reference I have included the tables, M, N, O, P, (demand at individual water outlets) and Q as in appendix (A) 2014 Illinois State Plumbing Code.

Section 890.TABLE M   Load Values Assigned to Fixtures

Fixture	Occupancy	Type of Supply Control	Load Values in Water (Supply Fixture Units)		
			Cold	Hot	Total
Water Closet	Public/Private	Flush Valve	10	–	10
Water Closet	Public/Private	Flush Tank	3	–	3
Urinal	Public	1" Flush Valve	10	–	10
Urinal	Public	¾" Flush Valve	5	–	5
Urinal	Public	Flush Tank	3	–	3
Lavatory	Public	Faucet	1	1	2
Bathtub	Public	Faucet	3	3	4
Shower Head	Public	Mixing Valve	2	2	3
Service Sink	Offices, etc.	Faucet	2	2	3
Kitchen Sink	Hotel/Restaurant	Faucet	3	3	4
Drinking Fountain	Office, etc.	3/8" Valve	0.25	–	0.25
Lavatory	Private	Faucet	0.75	0.75	1
Bathtub	Private	Faucet	1.5	1.5	2
Shower Stall	Private	Mixing Valve	1	1	2
Kitchen Sink	Private	Faucet	1.5	1.5	2
Laundry Trays (1 to 3)	Private	Faucet	2	2	3
Dishwashing Machine	Private	Automatic		-1	1
Laundry Machine (8 lb)	Private	Automatic	1.5	1.5	2

| Laundry Machine (16) | Public/ General | Automatic | 3 | 3 | 4 |

Section 890.TABLE N Water Supply Fixture Units (WSFU) for a Supply System with Flush Tanks Water Closets

WSFU	Demand (GPM)	Pipe Size (Inches)	Pressure Loss (PSI/100' of Pipe)	Velocity (Ft./Sec.)	Meter Size (Inches)
2	2	½"	4.2	2.7	⅝"
4	3	½"	8.7	4.2	⅝"
6	5	½"	22.5	7.0	⅝"
8	6.5	¾"	6.3	4.3	⅝"
10	8	¾"	9.0	5.4	¾"
12	9.2	¾"	11.5	6.1	¾"
14	10.4	¾"	15.0	6.9	¾"
16	11.6	¾"	18.0	7.7	¾"
20	14	1"	7.2	5.6	¾"
25	17	1"	10.0	6.6	¾"
30	20	1"	13.6	8.0	1"
35	22.5	1¼"	5.8	5.7	1"
40	25	1¼"	7.0	6.3	1"
45	27	1¼"	8.2	6.9	1"
50	29	1¼"	9.5	7.4	1"
60	32	1½"	5.0	5.8	1½"
70	35	1½"	6.2	6.4	1½"
80	38	1½"	7.0	7.2	1½"
90	41	1½"	8.0	7.5	1½"
100	43.5	1½"	8.7	7.8	2"
120	48	2"	2.7	5.0	2"
140	52.5	2"	3.1	5.4	2"
160	57	2"	3.6	5.8	2"
180	61	2"	3.9	6.1	2"
200	65	2"	4.5	6.6	2"
225	70	2"	5.2	7.1	2"

250	75	2"	6.0	7.7	3"
275	80	2½"	2.6	5.5	3"
300	85	2½"	2.9	5.8	3"
350	95	2½"	3.5	6.5	3"
400	105	2½"	4.2	7.1	3"
450	115	2½"	5.0	8.0	3"
500	125	3"	2.3	5.9	3"
600	145	3"	3.1	6.8	4"
750	170	3"	4.0	8.0	4"
1000	208	4"	1.5	5.7	4"

Section 890. TABLE O Water Supply Fixture Units (WSFU) for a Supply System with Flushometer Water Closets

WSFU	Demand (GPM)	Pipe Size (Inches)	Pressure Loss (PSI/100' of Pipe)	Velocity (Ft./Sec.)	Meter Size (Inches)
10	27	1¼"	8.3	6.8	¾"
12	28.6	1¼"	9.2	7.2	¾"
14	30.2	1¼"	10	7.9	¾"
16	31.8	1¼"	11	8.0	¾"
20	35	1½"	6.0	6.4	¾"
25	38	1½"	7.0	6.9	1"
30	41	1½"	8.0	7.4	1"
35	43.8	1½"	8.8	8.0	1"
40	46.5	2"	2.5	4.7	1"
45	49	2"	2.7	5.1	1"
50	51.5	2"	2.9	5.4	1½"
60	55	2"	3.4	5.8	1½"
70	58.5	2"	3.7	6.0	1½"
80	62	2"	4.0	6.2	1½"
90	64.8	2"	4.6	6.5	1½"
100	67.5	2"	5.0	6.8	1½"
120	72.5	2"	5.6	7.2	2"
140	77.5	2"	6.3	8.0	2"
160	82.5	2½"	2.7	5.7	2"
180	87	2½"	3.0	6.1	2"
200	91.5	2½"	3.4	6.4	2"
225	97	2½"	3.7	6.8	2"
250	101	2½"	4.0	7.1	3"
275	106	2½"	4.2	7.3	3"
300	110	2½"	4.6	7.6	3"
350	119	3"	2.1	5.5	3"
400	126	3"	2.3	5.9	3"
450	138	3"	2.7	6.3	3"

500	145	3"	3.0	6.8	3"	
600	160	3"	3.6	7.4	4"	
750	178	4"	1.1	4.7	4"	
1000	208	4"	1.5	5.6	4"	
1250	240	4"	1.9	6.4	4"	
1500	267	4"	2.3	7.0	4"	
1750	294	4"	2.8	7.8	4"	
2000	321	6"	0.4	3.7	6"	

Section 890.TABLE P Demand at Individual Water Outlets Demand at Individual Water Outlets

Type of Outlet	Demand (g.p.m.)
Ordinary Lavatory Faucet	2.0
Self-Closing Lavatory Faucet	2.5
Sink Faucet, ⅜" or ½"	4.5
Sink Faucet, ¾"	6.0
Bath Faucet, ½"	5.0
Shower Head, ½"	5.0
Laundry Faucet, ½"	5.0
Ballcock in Water Closet Flush Tank	3.0
1" Flush Valve (25 psi flow pressure)	35.0
1" Flush Valve (15 psi flow pressure)	27.0
¾" Flush Valve (15 psi flow pressure)	15.0
Drinking Fountain Jet	0.75
Dishwashing Machine (domestic)	4.0
Laundry Machine (8 to 16 pounds)	4.0
Aspirator (operating room or laboratory)	2.5

Section 890.TABLE Q Allowance in Equivalent Length of Pipe for Friction Loss in Valves and Fittings The following applies to all types of material approved for potable water distribution.

Equivalent Feet of Pipe for Various Pipe Sizes

Valve or Fitting	½"	¾"	1"	1¼"	1½"	2"	2½"	3"
45° ell (wrought)	0.5	0.5	1.0	1.0	2.0	2.0	3.0	4.0
90° ell (wrought)	0.5	1.0	1.0	2.0	2.0	2.0	2.0	3.0
Tee, Run (wrought)	0.5	0.5	0.5	0.5	1.0	1.0	2.0	–
Tee, Branch (wrought)	1.0	2.0	3.0	4.0	5.0	7.0	9.0	–
45° ell (cast)	0.5	1.0	2.0	2.0	3.0	5.0	8.0	11.0
90° ell (cast)	1.0	2.0	4.0	5.0	8.0	11.0	14.0	18.0
Tee, Run (cast)	0.5	0.5	0.5	1.5	1.0	2.0	2.0	2.0
Tee, Branch (cast)	2.0	3.0	5.0	7.0	9.0	12.0	16.0	20.0
Compression Stop	13.0	21.0	30.0	–	–	–	–	–
Globe Valve	–	–	–	53.0	66.0	90.0	–	–
Gate Valve	–	–	1.0	1.0	2.0	2.0	2.0	2.0

11 HOT WATER SUPPLY

Hot Water Introduction

The delivery of hot water to faucets, showers, clothes washers, dishwashers, and other hot water-using products in the typical home is an important driver in the water use profile of that home. In some homes, the design of the plumbing system is such that water is wasted while the user is waiting for hot water to arrive at the end fixture. This waste may be avoidable with a properly designed plumbing system. In new homes, the emphasis is upon "structured plumbing", while in older homes (where it is usually not cost-effective to install a new "structured plumbing" system), certain "add-on" pieces of equipment can sometimes be installed to reduce water waste.

Not all hot water tanks are created equal; however, all hot water tanks to be installed in Illinois shall meet the basic requirements as per ASHRAE 90 standards. (American Society of Heating and Air Conditioning Engineers). Though it is often confused as a plumbing fixture, a water heater falls under the category of appliance. On the other hand, more specifically a plumbing appliance.

This standard provides the minimum requirements for energy-efficient design of most buildings, except low-rise residential buildings. It offers the minimum energy-efficient requirements for design construction of new buildings and their systems, well as criteria for determining compliance with these requirements. It is an indispensable reference for engineers and other professionals involved in design of buildings. In addition to ASHRAE standards, water heaters must also meet the construction requirements of ASME (American Society of Mechanical Engineers), AGA (American Gas Association), or UL (Underwriters Laboratories).

Keep in mind that in Illinois under License Law, a licensed plumber is the only person that can and should install water heaters. The only exception is homeowners, as long as they have a permit and install the unit as per the plumbing code. That means handymen and HVAC (heating ventilation and air conditioning) companies are not allowed to install a water heater.

Types of Water Heaters

It is a good idea to know the different types of water heaters available, as well as how they function.

Conventional storage water heaters

Offer a ready reservoir (storage tank) of hot water. A single-family storage water heater offers a ready reservoir – from 20 to 80 gallons – of hot water. It operates by releasing hot water from the top of the tank when you turn on the hot water tap. To replace that hot water, cold water enters the bottom of the tank, ensuring that the tank is always full. Since water is constantly heated in the tank, energy can be wasted even when a hot water tap is not running. This is called *standby heat loss*. Only tankless water heaters – such as demand-type water heaters and tankless coil water heaters – avoid standby heat losses. Some storage water heater models have heavily insulated tank, which significantly reduce standby heat losses and lower annual operating costs.

Routine maintenance for storage water heaters, depending on what type or model you have may include:

Flushing a quart of water from the storage tank every three months Checking the temperature and pressure valve every six months Inspecting the anode rod every three to four years.

Tankless or demand-type water heaters

Heat water directly without the use of a storage tank

Tankless water heaters, also known as demand-type or instantaneous water heaters, provide hot water only as needed. They do not produce the standby energy loss associated with storage water heaters, which can save money. Tankless water heaters heat water directly without the use of a storage tank. When a hot water tap is turned on, cold water travels through a pipe into the unit. Either a gas burner or an electric element heats the water. As a result, tankless water heaters deliver a constant supply of hot water. You do not need to wait for a storage tank to fill up with enough hot water. However, a tankless water heater's output limits the flow rate. Typically, tankless water heaters provide hot water at a rate of 2–5 gallons per minute. Gas-fired tankless water heaters produce higher flow rates than electric ones. Sometimes even the largest, gas-fired model cannot supply enough hot water for simultaneous, multiple uses in large households. For example, taking a shower and running the dishwasher at the same time can stretch a tankless water heater to its limit. To overcome this problem, you can install two or more tankless water heaters, connected in parallel for simultaneous demands of hot water. For homes that use 41 gallons or less of hot water daily, demand water heaters can be 24%–34% more energy efficient than conventional storage tank water heaters.

They can be 8%–14% more energy efficient for homes that use a lot of hot water – around 86 gallons per day. You can achieve even greater energy savings of 27%–50% if you install a demand water heater at each hot water outlet. Energy Star estimates that a typical family can save $100 or more per year with an Energy Star qualified tankless water heater. The initial cost of a tankless water heater is greater than that of a conventional storage water heater, but tankless water heaters will typically last longer and have lower operating and energy cost. Which could offset its higher purchase price. Most tankless water heaters have a life expectancy of more than 20 years. They also have easily replaceable parts that extend their life by many more years. In contrast, storage water heaters last 10–15 years.

Tankless water heaters can avoid the standby heat losses associated with storage water heaters. Although gas-fired tankless water heaters tend to have higher flow rates than electric ones, they can waste energy if they have a constantly burning pilot light. This can sometimes offset the elimination of standby energy losses when compared to a storage water heater. In a gas-fired storage water heater, the pilot light heats the water in the tank so the energy is not wasted. The cost of operating a pilot light in a tankless water heater varies from model to model. Ask the manufacturer how much gas the pilot light uses for the model you are considering. If you purchase a model that uses a standing pilot light, you can always turn it off when it is not in use to save energy. Also, consider models that have an intermittent ignition device (IID) instead of a standing pilot light. This device resembles the spark ignition device on some gas kitchen ranges and ovens.

Tankless or demand-type water heaters are rated by the maximum temperature rise possible at a given flow rate. Therefore, to size a demand water heater, you need to determine the flow rate and the temperature rise you will need for its application (whole house or a remote application, such as just a bathroom) in your home. First, list the number of hot water devices you expect to use at any one time. Then, add up their flow rates (gallons per minute). You will want for the demand water heater this desired flow rate. For example, let us say you expect to simultaneously run a hot water faucet with a flow rate of 0.75 gallons per minute and a showerhead with a flow rate of 2.5 gallon per minute. The flow rate through the demand water heater would need to be at least 3.25 gallons per minute. To reduce flow rates, install low-flow water fixtures.

To determine temperature rise, subtract the incoming water temperature from the desired output temperature. Unless you know otherwise, assume that the incoming water temperature is 50°F. For most uses, you will want your water heated to 120°F. In this example, you would need a demand water heater that produces a temperature rise of 70°F for most uses.

For dishwashers without internal heaters and other such applications, you might want your water heated at 140°F. In that case, you will need a temperature rise of 90°F.

Most demand water heaters are rated for a variety of inlet temperatures. Typically, a 70°F water temperature rise is possible at a flow rate of 5 gallons per minute through gas-fired demand water heaters and 2 gallons per minute through electric ones. Faster flow rates or cooler inlet temperatures can sometimes reduce the water temperature at the most distant faucet. Some types of tankless water heaters are thermostatically controlled; they can vary their output temperature according to the water flow rate and inlet temperature.

Heat pump water heaters

Move heat from one place to another instead of generating heat directly for providing hot water.

Heat pump water heaters use electricity to move heat from one place to another instead of generating heat directly. Therefore, they can be two to three times more energy efficient than conventional electric resistance water heaters. To move the heat, heat pumps work like a refrigerator in reverse. While a refrigerator pulls heat from inside a box and dumps it into the surrounding room, a stand-alone *air-source heat pump* water heater pulls heat from the surrounding air and dumps it – at a higher temperature –

Into a tank to heat, water. You can purchase a stand-alone heat pump water heating system as an integrated unit with a built-in water storage tank and back-up resistance heating elements. You can also retrofit a heat pump to work with an existing conventional storage water heater.

Heat pump water heaters require installation in locations that remain in the 40°–90°F range year-round and provide at least 1,000 cubic feet of air space around the water heater. Cool exhaust air can be exhausted to the room or outdoors. Install them in a space with excess heat, such as a furnace room. Heat pump water heaters will not operate efficiently in a cold space. They tend to cool the spaces that they are placed in. You can also install an air-source heat pump system that combines heating, cooling, and water heating. These combination systems pull their heat indoors from the outdoor air in the winter and from the indoor air in the summer. Because they remove heat from the air, any type of air-source heat pump system works more efficiently in a warm climate. Homeowners primarily install geothermal heat pumps, which draw heat from the ground during the winter and from the indoor air during the summer – for heating and cooling their homes. For water heating, you can add a *desuperheater* to a geothermal heat pump system.

A desuperheater is a small, auxiliary heat exchanger that uses superheated gases from the heat pump's compressor to heat water. This hot water then circulates through a pipe to the home's water heater tank. Desuperheaters are also available for tankless or demand- type water heaters. In the summer, the desuperheater uses the excess heat that would otherwise be expelled to the ground. Therefore, when the geothermal heat pump runs frequently during the summer, it can heat all of your water.

During the fall, winter, and spring- when the desuperheater is not producing as much excess heat- you will need to rely more on your storage or demand water heater to heat the water. Some manufacturers also offer triple-function geothermal heat pump systems, which provide heating, cooling, and hot water. They use a separate heat exchanger to meet all of a household's hot water needs.

Heat pump water heater systems typically have higher initial costs than conventional storage water heaters. However, they have lower operating costs, which can offset their higher purchase and installation prices. In order to size a water heater properly for your home, including a heat pump water heater with a tank, use the water heater's first hour rating. The first hour rating is the number of gallons of hot water the heater can supply per hour (starting with a tank full of hot water). It depends on the tank capacity, source of heat (burner or element), and the size of the burner or element.

Solar water heaters

Use the sun's heat to provide hot water.
Solar water heating systems include storage tanks and solar collectors. There are two types of solar water heating systems: active, which have circulating pumps and controls, and passive, which do not.
Active Solar;
There are two types of active solar water heating systems:

Direct circulation systems
Pumps circulate household water through the collectors and into the home. They work well in climates where it rarely freezes.

Indirect circulation systems
Pumps circulate a non-freezing, heat-transfer fluid through the collectors and a heat exchanger. This heats the water that then flows into the home. They are popular in climates prone to freezing temperatures.

Passive Solar;
Passive solar water heating systems are typically less expensive than active systems, but they are usually not as efficient. However, passive systems can be more reliable and may last longer. There are two basic types of passive systems:

Integral collector-storage passive systems
These work best in areas where temperatures rarely fall below freezing. They also work well in households with significant daytime and evening hot-water needs.

Thermosyphon systems

Water flows through the system when warm water rises as cooler water sinks. The collector must be installed below the storage tank so that warm water will rise into the tank. These systems are reliable, but contractors must pay careful attention to the roof design because of the heavy storage tank. They are usually more expensive than integral collector-storage passive systems. Most solar water heaters require a well-insulated storage tank. Solar storage tanks have an additional outlet and inlet connected to and from the collector. In two-tank systems, the solar water heater preheats water before it enters the conventional water heater. In one-tank systems, the back-up heater is combined with the solar storage in one tank. Solar water heating systems usually require a backup system for cloudy days. Moreover, times of increased demand. Conventional storage water heaters usually provide backup and may already be part of the solar system package. A backup system may also be part of the solar collector, such as rooftop tanks with thermosyphon systems. Since an integral-collector storage system already stores hot water in addition to collecting solar heat, it may be packaged with a tankless or demand-type water heater for backup.

Tankless coil and indirect water heaters

Tankless coil and indirect water heaters use a home's space heating system to heat water. They are part of what is called integrated or combination water and space heating systems. A tankless coil water heater provides hot water on demand without a tank. When a hot water faucet is turned on, water is heated as it flows through a heating coil or heat exchanger installed in a main furnace or boiler. Tankless coil water heaters are most efficient during cold months when the heating system is used regularly but can be an inefficient choice for many homes, especially for those in warmer climates. Indirect water heaters are a more efficient choice for most homes, even though they require a storage tank. An indirect water heater uses the main furnace or boiler to heat a fluid that is circulated through a heat exchanger in the storage tank. The energy stored by the water tank allows the furnace to turn off and on less often, which saves energy.
An indirect water heater, if used with a high-efficiency boiler and well- insulated tank, can be the least expensive means of providing hot water, particularly if the heat source boiler is set to "cold start."

Piping Methods

To parallel or to series has been the debate for as long as I can remember, with both sides arguing over which is the best method for piping two water heaters. In situations where hot water volume requirements are frequently in high demand, parallel may be the best solution.

When water heaters are connected in series, it means the cold water is fed through each tank, one after the other, with the first tank's hot outlet becoming the next tanks cold inlet. Under normal conditions, the first tank in the series will do most of the heating. Under high demand situations, the first tank will not be able to sufficiently heat the water so the next tank in series would begin operation, and the next...until the water is up to temperature.
Points to consider when connecting water heaters in series are:
- A bypass must be installed so that any malfunctioning tank can be removed while keeping the system operational.
- Different styles / models of water heaters can be used in series but the tank with the highest BTU rating should be fed first.
- Pipe size is limited by the tanks inlet and outlet.
- The system operates efficiently under both low and peek demands.
- No requirement to balance the piping as with a parallel installation.

Large volumes of water are heated and ready for the demand. Conversely, if hot water demand were only occasionally peaking, a series configuration would allow you to turn on/off the upstream heater at will, heating additional water when needed, and then temporarily disabling the tank, thus saving energy, when the demand is not there. It is important to mention, turning off the power source (not the supply) to the upstream heater will continue to give you hot water on demand from the downstream heater. However, vice versa would result in having to draw all the water from the downstream heater before the heated water at the upstream heater would be available to the demand point. Another reason for installing a second water heater in series is you have already determined you want more hot water available, but you are not ready to trash your current water heater. Adding another heater in series will allow you to use both heaters with doubled capacity. Usually a cheaper alternative than buying one water heater twice the size of the original.

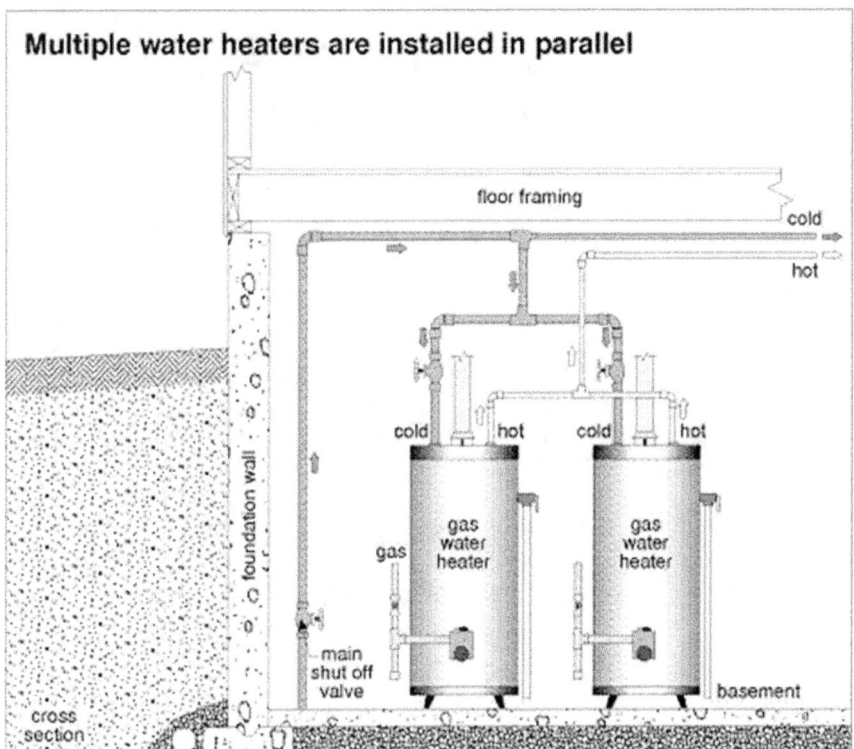

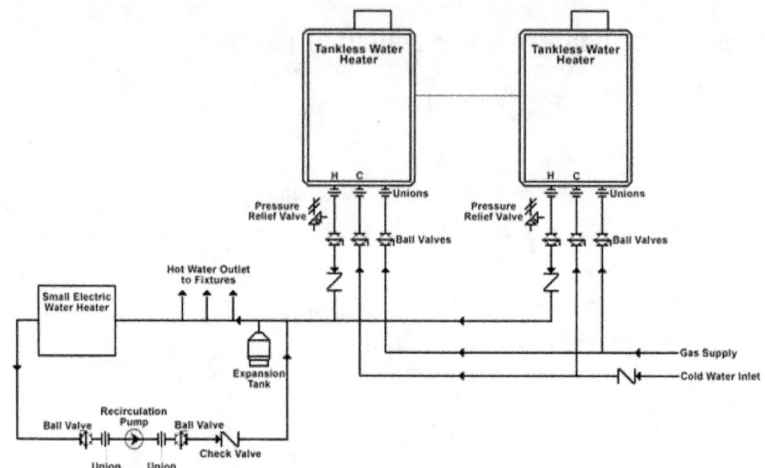

Tankless water heaters piped in series

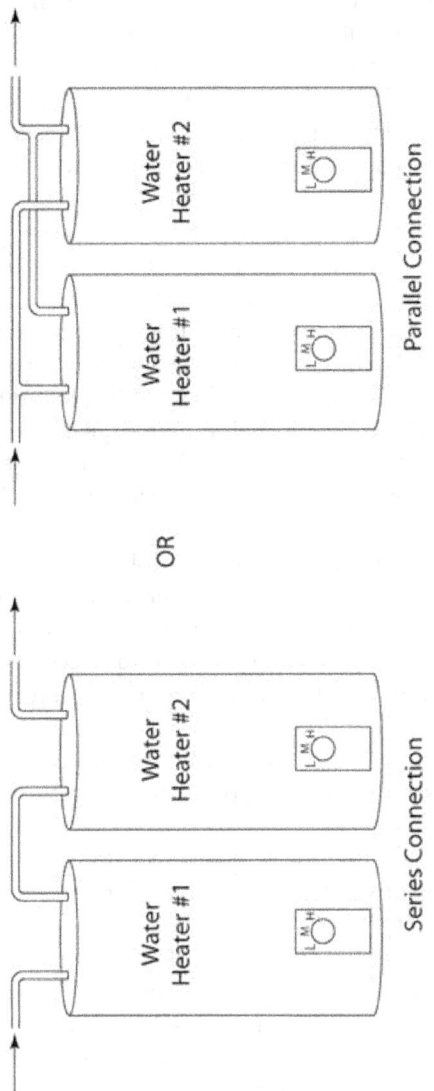

Water Heater Maintenance

Maintenance on water heaters will vary, depending on the type. It is equally important to know that all water heaters require some maintenance to ensure a longer efficient life of the unit. With that, water quality is a major part of how long a water heater will last. If there is one question most commonly asked by a homeowner it is "how long should my water heater last?" which can be a difficult question to answer. First, let us discuss the two different water sources.

Water source number one being of a public water system or provided by a municipality. This, as we now know from previous chapters, is always treated water. The fact is, with the exception of Chicago water, these sources of water are often from a well. The next source of water I would like to bring up is that of a privately owned well. This is most common in rural areas where water is not supplied by a municipality. First, I would like to point out that well drilling is not plumbing. Without going into a three-page dissertation as to what the legal description of exactly what a well driller is, and responsibilities are, let us keep it simple. A well driller is a contractor licensed by the Illinois Department of Public Health.

The short version of this description is as follows.

Subject to the provisions of Section 3, after January 1, 1972 no person shall drill, install or repair a water well pump or equipment, or engage in the occupation of a water well and pump installation contractor unless he holds a valid license as a water and well and pump installation contractor issued under this act.

Anytime a new well is drilled and installed, the water is to be sampled and tested to be sure it is safe to drink, prior to use. Some of what the sample will reveal and is tested for are:

Hardness: the hardness of water is a measure of the amount of minerals, primarily calcium and magnesium it contains. Adding a water softener will help reduce the amount of hardness if needed. A few other things that well water is tested for include sulfates. Once again, this means in groundwater sulfates are caused by natural deposits of magnesium sulfate, calcium, or sodium. Concentrations should be below 250 ppm (parts per million). Higher concentrations are undesirable because of their laxative effects. Iron in drinking water can be objectionable because it can give a rusty color to laundry, and may affect the taste. Chlorides in groundwater can be naturally occurring in deep aquafers or caused by pollution from seawater, brine, or industrial or domestic waste.

PH, (potential hydrogen) is a measure of the acid or alkaline content of water, PH values range from 0-14. The lower the PH value the more acid the water is. The higher the PH value the more alkaline the water. PH for drinking water normally ranges from 5.5 to 9.0. At PH less than 7.0, corrosion of water pipes may occur.

Total dissolved solids test measures the total amount of dissolved minerals in water. The solids can be iron, chlorides, sulfates, calcium or other minerals found on the earth's surface. The dissolved minerals can produce an unpleasant taste or appearance and can contribute to scale deposits on pipe walls.

Alkalinity is a measure of the presence of bicarbonate, carbonate, or hydroxide constituents. Concentrations of less than 100 ppm are desirable for domestic water supplies. The recommended range for drinking water is 30 to 400 ppm. A minimum level of alkalinity is desirable because it is considered a "buffer" that prevents large variations in PH.

Alkalinity is not detrimental to humans. Moderately alkaline water (less than 350 mg/l); in combination with hardness, forms a layer of calcium or magnesium carbonate that tends to inhibit corrosion of metal piping. Many public water utilities employ this practice to reduce pipe corrosion and to increase the useful life of the water distribution system.

High alkalinity (above 500 mg/l) is usually associated with high PH values, hardness, and high dissolved solids. It has adverse effects on plumbing systems, especially on hot water systems (water heaters, boilers, heat exchangers, etc.). Excessive scale reduces the transfer of heat to the water, thereby resulting in greater power consumption and increased costs. Water with low alkalinity (less than 75 mg/l), especially some surface waters and rainfall, is subject to changes in PH due to dissolved gasses that may be corrosive to metallic fittings.

Tastes and odors in water may be caused by hydrogen sulfide (H2S). Hydrogen sulfide, when dissolved in water, produces an offensive odor resembling that of rotten eggs. The presence of hydrogen sulfide in deep well water is due to the reduction of sulfate. The acceptable level of hydrogen sulfide is 0.05 mg/l or less. Hydrogen sulfide can be removed through oxidation, by aeration, or chlorination. The precipitated sulfur should be removed by filtration to prevent it from reverting to hydrogen sulfide through the action of certain microorganisms.

The oxidation of hydrogen sulfide by chlorine may be advantageous in cases where it is otherwise unnecessary to re-pump the water (which is normally required with aeration), because chlorine can be applied directly into the system. Enough chlorine must be used to maintain a distinct chlorine residual. The anode rod in the water heater is a huge part of required maintenance. As to exactly what is an anode rod, and what its function is? It is a sacrificial rod to protect the water heater itself. Replacing the anode rod before it fails can slow down corrosion inside the tank, and prolong the life of the unit. What or how is this related to well water? The simple reason is most well water systems are left untreated, to the same level as that in a municipality water source. About 12 million American households, roughly 15 percent of the U.S. population, draw their drinking water from private wells. Unlike public water systems, which are regulated by the U.S. Environmental Protection Agency (EPA), private wells are the responsibilities of homeowners. EPA standards do not apply to private wells.

Consequently, well-owning homeowners are obliged to protect and maintain their water supplies to optimize the quality of the drinking water supplied to their families. Many factors contribute to the quality of private well water. Some, such as routine testing and treatment, and properly positioning wells relative to point sources of contamination (including in-ground septic systems), are within the control of homeowners. Others, such as regional ground water quality and flooding episodes, are much less so. For the private well owner, good quality drinking water depends on a multi-barrier approach to contamination that includes well monitoring and maintenance.

Ground water is not 100 percent pure water. Because it collects in the tiny pore spaces within sediments and in the fractures within bedrock, ground water always contains some dissolved minerals. In addition, because there is some life form occupying virtually every geological niche, there are many naturally occurring microorganisms in ground water.

This, along with a magnesium anode rod, will cause a rotten egg smell from the hot water tap- in some cases almost immediately after the installation of a new water heater.

Once the anode rod is removed, the manufactures warranty becomes void. The alternative and in most instances the solution to the problem is to include, at the time of install, an upgraded anode rod. Remember, the anode rod is a sacrificial rod designed to fail. Without one, the tank will also become sacrificial, and will fail at a much higher rate.

The options are; the installation of an aluminum zinc anode rod or the installation of a powered anode rod. This device is sold complete with a transformer and can be installed in place of the standard anode rod. Voltage ratings can be adjusted to accommodate simple or unusual water conditions. A stabilizing weight prevents the anode from shorting out against the tank and serves as a dielectric standoff to prevent localized over protection at the tank bottom.

The special ceramic anode design provides an anode design of one century for glass lined tanks. The impressed current anode system has a life many times that of a sacrificial anode, even in corrosive water.

The combined presence of hydrogen, sulfur, and bacteria cause foul rotten egg smell as well. The magnesium anode will protect the tank surface but generates enough hydrogen to create an odor when it interacts with sulfur in the water or bacteria in the tank.

I would like to point out one other interesting note, under normal circumstances, rainwater and melted snow trickle gradually into the ground through the tiny spaces between grains of sediment. This action results in the natural filtration of ground water, in which particles, even bacteria, are separated out of the water by a "sieve effect." During periods of flooding however, natural filtration is bypassed and wells can become contaminated rapidly. Shallow wells are at greater risk for contamination than deep wells during floods. According to the EPA, wells that are more than 10 years old or less than 50 feet deep are likely to be contaminated following a flood, even if there is no apparent damage.

If microbial contamination is discovered in private well water, immediate disinfection is required. This task can be carried out either by ground water professionals or by the homeowner using an array of information resources available from state and local health departments and government agencies. The most commonly used well water disinfectants are sodium hypochlorite (chlorine bleach) and calcium hypochlorite. Word of caution, ONLY a licensed professional should administer or treat a private well system. Over chlorination is just as harmful as the bacteria itself.

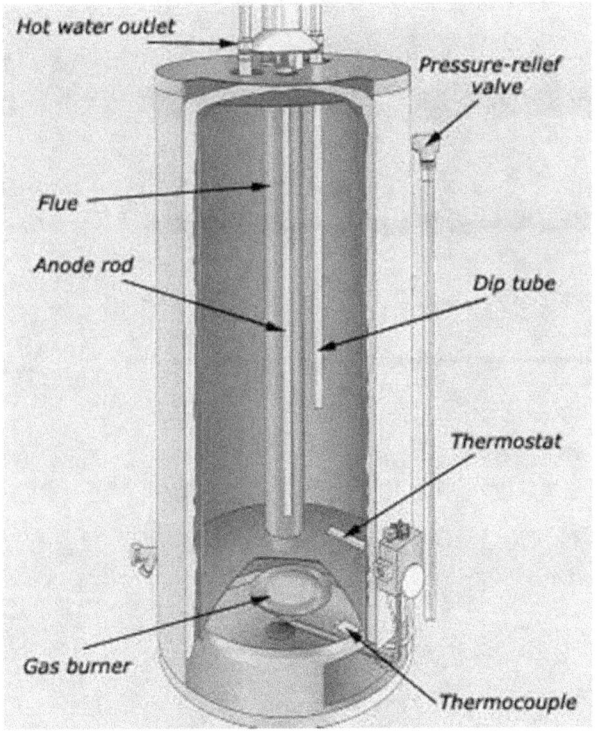

Maintaining a water heater on a regular basis will help to improve customer relations and provide an annual inspection of the unit to ensure your customers will have peace of mind. Let us be honest: how many homeowners are routinely maintaining their water heater? One of the first things, especially if you are providing service to a new customer, is to check the set temperature. Ensure the temperature is within a safe setting. Second, as recommended by manufactures, the water heater should be drained annually. This will help flush mineral deposits from the tank. Third, the anode rod should be removed and inspected for deterioration.
As always, if you are uncertain where the location of the anode rod is check the installation guide. (Usually in the plastic sleeve stuck to the side of the heater.)

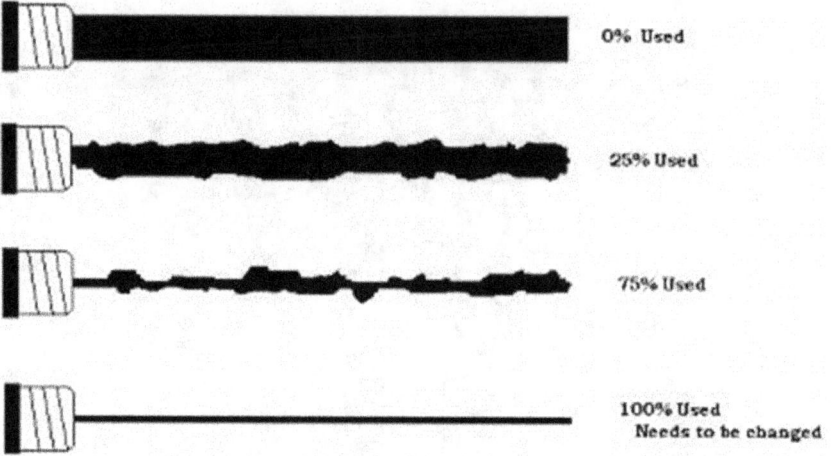

Anode rod in various stages of wear

12 DRAIN WASTE AND VENT SYSTEM

Introduction

The drainage system is a series of pipes, which receives, carries, and removes waste and rainwater, other liquids, and human excreta from fixtures to a sewer or other disposal receptacle. This system is divided into several sections, each of which is defined below.

Parts of the drainage system

House sewer: The section of pipe, which runs between the house drainage system and the connection to the public sewer or septic tank. House sewers should convey the waste of only one residence.

House drain: The lowest piping in a house drainage system, this pipe receives the discharge from soil, waste, and other drainage pipes, and then carries such discharge to the house sewer. The house drain ends just outside the front or foundation wall of the building, and operates by gravity.

Soil stack and pipe: Any line of pipe, which carries the discharge of water closets. The term "stack" refers to the vertical runs of such piping.

Waste stack and pipe: All pipe receiving the discharge of fixtures other than water closets. An indirect waste pipe does not connect directly with either the house drain or the soil or waste stack, but usually ends over and above the overflow rim of fixture that is water- supplied, trapped, and vented.

Trap: Refers to a fitting or device constructed to prevent the passage of air or gas back through a pipe or fixture, without materially affecting the flow of sewage or wastewater.

Vent piping: Provides ventilation to the drainage system and prevents trap siphonage and backpressure from clogging or contaminating the drainage system. Local ventilating pipe is a duct or pipe connected to the house side of a fixture or trap through which foul vapors may be removed from a room.

Fixture: Any receptacle intended to receive or discharge water or water-carried waste into the drainage system.

Main: Any system of horizontal, vertical, or continuous piping to which fixtures are connected either directly or with branches.

Branch: That part of the plumbing system, which extends from the main to a fixture.

Leader: Any vertical line of piping which receives and carries rainwater.

Fitting: Any one of a number of devices used to connect pieces of pipe or change the direction of pipe.

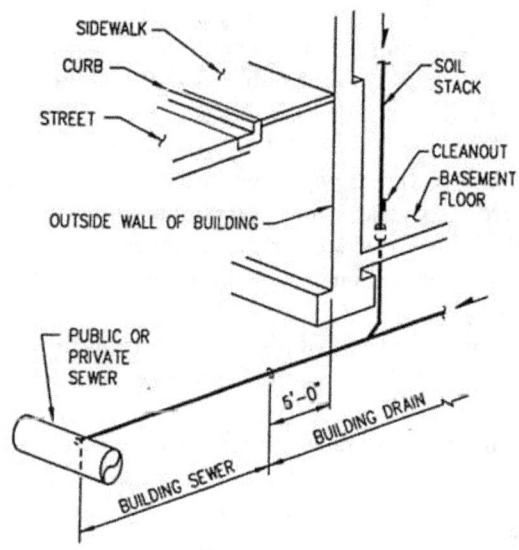

Drain stacks in older homes are often made of cast iron, which rusts through after 80 years or so. In older homes, the branch drains typically are made of galvanized steel, which is much more likely to rust and corrode shut. In newer homes, plastic pipe is used for stacks and branch drains. The first plastic pipe to be used was ABS, which is black. Since the 1970s, ABS has generally been replaced by PVC pipe. In some residential construction, it was also common practice to utilize copper, for use in drain waste and vent.

There are three types of drainage systems. This includes sanitary, storm, and condensate. Within each of the three, there are sub categories of drainage. Within sanitary there is only two categories to consider, fecal matter and grey water waste such as laundry waste.

Within the storm drainage system, there are sub-soil drainage, draintile, and surface run-off. Within the condensate drainage system, such as an air conditioner, discharge from the temperature and pressure valve on the water heater, condensate from a commercial cooler, or almost any other clear water waste you can think of.

Then there is also pumping equipment, which is also part of the drainage system. First, a look at subsoil drainage and the types that are most common. A subsoil drainage system is designed to be installed within the ground in order to remove excessive water from the soil, so that it does not cause damage to buildings or to the landscape. Because they are buried beneath the soil, they should be planned during the landscaping or building stage in order to get the best possible results.

The most common type of subsoil drainage is in the form of a special pipe that is covered with a geotextile material. The pipe itself has a number of perforations through it to let the water through and be drained away. The geotextile covering is important, as it stops soil and other particles from going through to the pipe and clogging it up so that it cannot serve its purpose of draining the water away. This is most referred to as perforated pipe with sock. Another type of subsoil drainage that you can utilize is to bury some rocks or pebbles. These are free draining and will let water through quite readily.

A French drain collects water from a soggy area and distributes that water to drier areas. French drain systems often contain one or more pipes buried in the gravel trench to speed up the water flow. Remember that there are two types of drainpipe: perforated and solid. Perforated pipe has many holes punched or drilled into it to allow water to both enter and exit the pipe. In any drainage system design, there is often water flowing IN through the perforations in a soggy area at the same time there is water flowing OUT of the perforations in a drier area. The purpose of a drywell is to allow water to enter the deeper subsoil faster and easier. It is simply a hole dug downward into the subsoil and filled with gravel or a sleeve. A small drywell can be dug with a posthole digger. A large drywell may use a drum or a precast concrete cylinder as a sleeve.

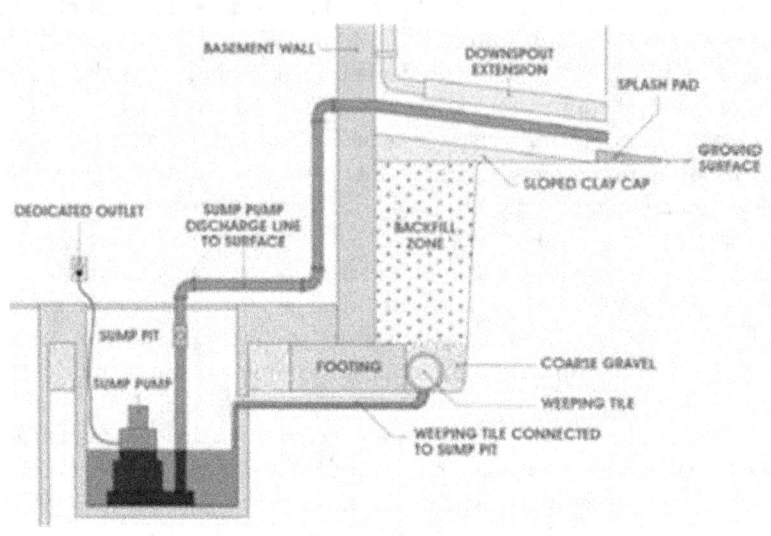

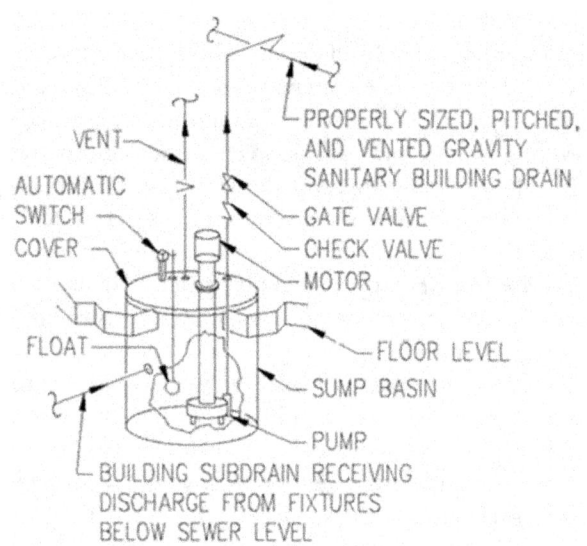

Residential Drainage

Section 890.170 Sewer and/or Water Required
Each building, which is intended for human habitation or occupancy, shall have a connection to a public water system, a semi-private water system, or a private water supply. In addition, a connection to a public sewer system or private sewage disposal system. In the diagram below, we can see this is a typical public sewer system. In addition, we can also determine this is an example of a gravity sewer.

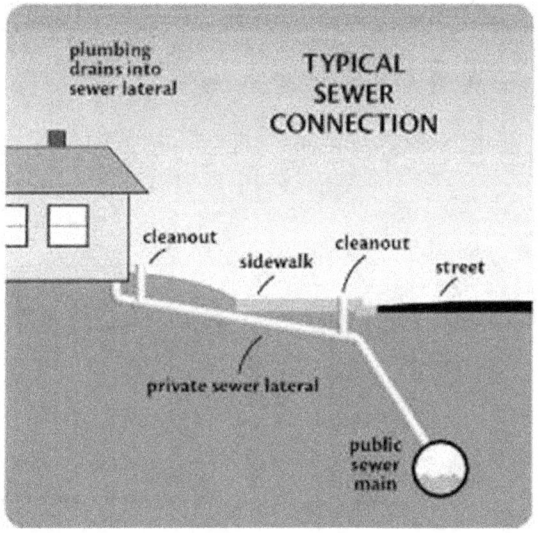

There are two types of sanitary sewers in a residential home. The first as listed above is that of a gravity sewer. The second is known as an overhead sewer.

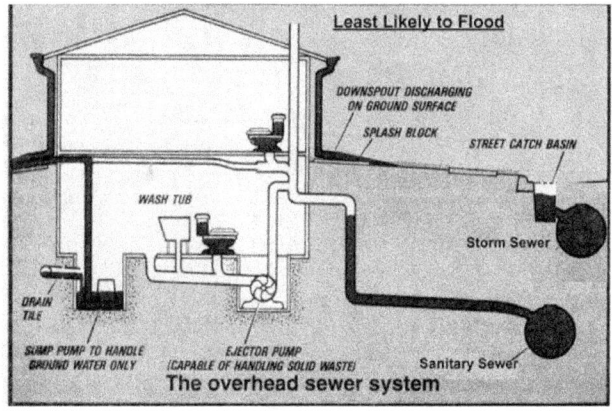

The system that is least likely to flood is the overhead sewer system. The reason that a gravity sewer may cause flooding in a home is that during heavy rains, the sewer system may become overwhelmed with the amount of water flowing into it.

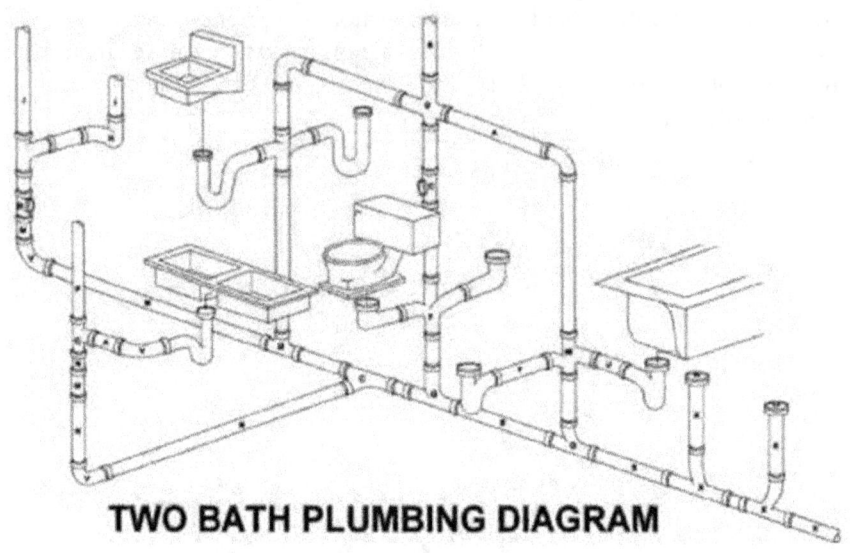

TWO BATH PLUMBING DIAGRAM

Horizontal to Vertical
Fittings A through E may be used for vertical drain piping picking up horizontal branches.
Fittings F to J may be used for horizontal drain piping changing to a vertical direction.

Vertical to Horizontal
Fittings A, B, C, F, G and& J may be used for this type of change in direction. Fitting J may be used when installed in a true vertical position.

Horizontal to Horizontal
Fittings A, B, C, F, G and I may be used for this type of change in direction.

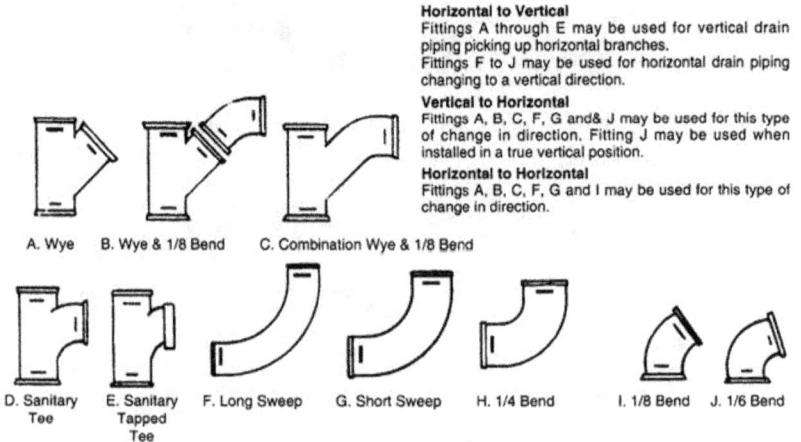

A. Wye   B. Wye & 1/8 Bend   C. Combination Wye & 1/8 Bend
D. Sanitary Tee   E. Sanitary Tapped Tee   F. Long Sweep   G. Short Sweep   H. 1/4 Bend   I. 1/8 Bend   J. 1/6 Bend

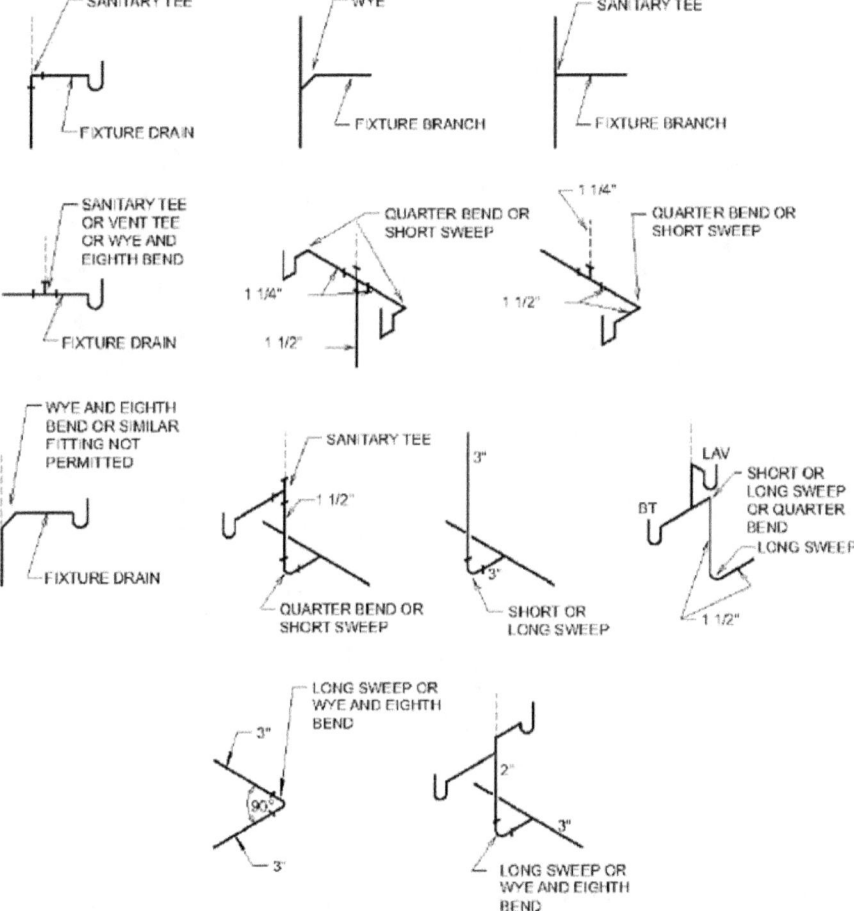

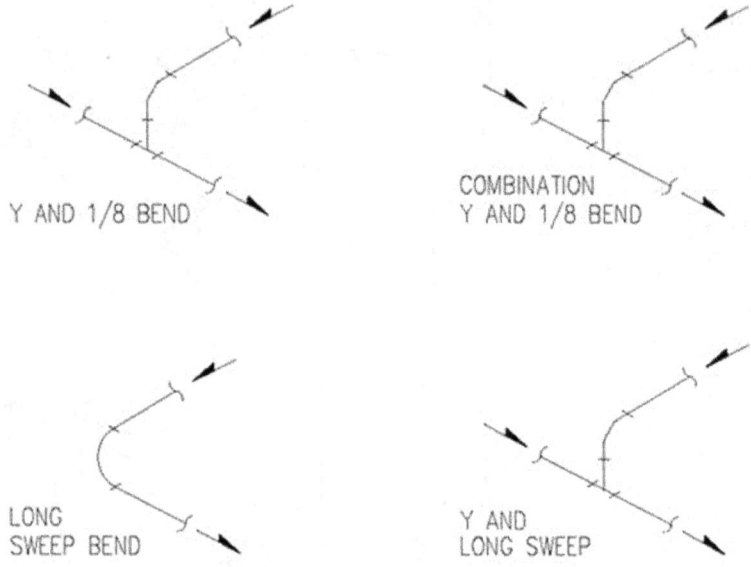

Another helpful tool for new apprentices just starting out in the trade is to get a copy of the Charlotte Pipe Company's fitting book. Their book contains every fitting they offer as well as the dimensions for each. The Charlotte fitting book is small enough to fit in your pocket or toolbox. You can get it simply by logging onto their website. I have provided the information in the back of this book, along with several other websites.

Drainage Fixture Units

As described in the definitions section of this book, a drainage fixture unit, or (DFU) as it is commonly known, is the value or the calculated total load carried by a soil or waste pipe. Each fixture will put a load on the waste pipe, and in order to properly design a waste system this formula is an important piece of the puzzle. In other words, each fixture that discharges into a waste pipe has a calculated load value. Now for the exception to the rule, in order to calculate a continuous flow rate such as that of a sump pump or ejector pump there is a DFU load of two per each gallon of water used per minute. Let us first take a closer look at just what this means when calculating a vent for an ejector pump. The industry standard has always been that a two-inch vent was the applicable size vent to install. However, if we are to use the calculation as outlined in the state codebook we will see that a two vent is undersized.

In order to conclude this, we will need a pump to use as an example. Let us say we have a job that has specified the use of a Zoeller model 270 to lift waste from a basin and into an overhead sewer system.

If we look at the data sheet for this particular pump, which I have included below.

Zoeller Model 270

Specifications:
Motor Characteristics Motor 1 HP
Voltage 115 or 230 V
Phase 1 Ph
Hertz 60 Hz
RPM 3450 RPM
Type Permanent split capacitor Insulation Class B
Amps 7.5 – 15.0 Amps
Pump Characteristics Operation Automatic or non-automatic
Discharge Size 2" NPT
Solids handling 2" (50 mm) spherical solids Cord Length 20' (6 m) standard
Cord Type UL listed, 3-wire neoprene cord and plug Max Head 29' (8.8 m)
Max Flow Rate 132 GPM (500 LPM) Max Operating Temp 130° F (54° C)
Cooling Oil filled
Motor Protection Auto reset thermal overload

As you can see the maximum flow rate as listed by, the manufacture is 132 GPM (gallons per minute) if we refer back to our codebook under section 890.1330 (b) the codebook reads as,

Values for Continuous Flow. For a continuous or semi-continuous flow into a drainage system, such as from a pump, ejector, air- conditioning equipment, or similar devices, two DFUs shall be considered to be equal to each gallon per minute (gpm) of flow. Its math time, so let us calculate this and see what we come up with.

132 (gpm) x 2 = 264 this shows us that 264 is our total DFU (at maximum flow) if we then refer back to our codebook under Appendix (A) Table (H) (fig.7) as provided below, we can see that our flow rate of 264 would require a vent of huge proportions, and would need to be four inches in diameter. Now for the interesting part of this formula and why it is important to fully know what the equipment can and will do at various measurements.

Section 890.TABLE H   Horizontal Fixture Branches and Stacks

Maximum Number of Drainage Fixture Units (D.F.U.)
That May be Connected to:

Diameter of Pipe (inches)	Any Horizontal Fixture Branch	One Stack of 3 Stories in Height or 3 Intervals	More than 3 Stories in Height — Total for Stack	More than 3 Stories in Height — Total at One Story or Branch Interval
1¼	1	2	2	1
1½	3	4	8	2
2	6	10	24	6
2½	12	20	42	9
3	20[1]	30[2]	60[2]	16[1]
4	100	240	500	90
5	360	540	1,100	200
6	620	960	1,900	350
8	1,400	2,200	3,600	600
10	2,500	3,800	5,600	1,500
12	3,900	6,000	8,400	1,500
15	7,000	–	–	–

[1] Not over two water closets.
[2] Not over six water closets or more than two per branch interval or per floor.

Figure 7

Using the chart in (fig. 8), we can see what the pump can actually produce in relationship to a given distance.

MODEL		270/4270	
Feet	Meters	Gal.	Liters
5	1.5	132	500
10	3.0	101	382
15	4.6	77	291
20	6.1	56	212
25	7.6	29	110
Shut-off Head:		29 ft.(8.8m)	

Figure 8

This chart (fig. 8) from the same manufacture and the same model pump. As you can see, the first five feet of the piping connected to this pump can produce up to a 132 GPM. However, at a distance of 25 feet that flow rate is drastically decreased to 29 gallons per minute. Once we look back at our codebook, again (fig. 7) a two and a half inch pipe would be just fine for this particular pump. The bottom line is the more familiar we get with the products that we will be installing the better the installation will be, and the better our install will comply with plumbing code.

This is of course all arguable, but at the end of the day, in most installations, the plumbing inspector will ultimately approve a two-inch vent for this application. Concerning drainage in Illinois and on a national level, "currently" the minimum size allowed for use in the building drain portion of the waste system is four inch. I use the term "currently" for the simple reason that the four-inch building drain may change to three inch in the future. As long as water closet, continue to use less and less water for their operation.

(EPA act of 1992 standards for toilets & faucets)

Fixture	Current standard (1992)
Residential toilet	1.6 gpf
Residential faucet	2.2 gpm @ 60 psi
Residential showerhead	2.5 gpm @ 80 psi

The reason that this is being considered is that as the water consumption has decreased in water closets from the old 1.6 gallon per flush, to the possible 1-gallon, or even a .5-gallon per flush toilet.

There is an issue with solids within the waste piping becoming "left behind" in the pipe, as the water trickles by. If this is an addendum to the current codebook, this will in turn effect the change in a few other codes as well. One example is that we are currently allowed to have two water closets on a three-inch waste. This would most likely have to be amended for the reason most homes built today have at least three water closets in them. This is not necessarily a bad thing. A smaller diameter pipe will increase velocity, thereby helping to carry the waste downstream. Service plumbers are all too familiar with this issue, as the water consumption changes more calls for sewer back-ups occur. However, this can also lead to unwanted callbacks as well.

Perhaps one solution is to install pressure-assisted water closets, or low consumption flushometer, which may one day, be the normal fixture in all water closets, doing away with the tank style toilet forever. I have heard many complaints since 1992, regarding the flushing ability of low consumption water closets.

Hydraulic Gradient

By definition (Merriam-Webster dictionary), a line joining the points of highest elevation of water in a series of vertical open pipes rising from a pipeline in which water flows under pressure.

What does that mean? Simply put, this is the amount of fall or downhill slope in a given pipe or trench. The theory is too much pitch water will out run solids, too little pitch nothing will flow out of the pipe.

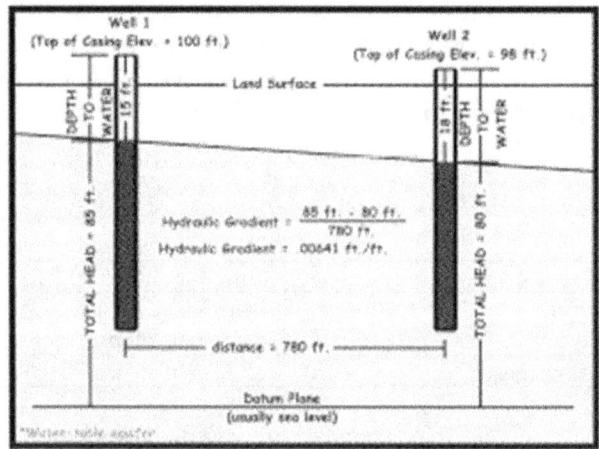

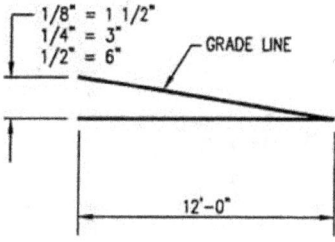

Illustration O (grade)

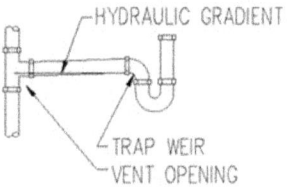

SATISFACTORY INSTALLATION

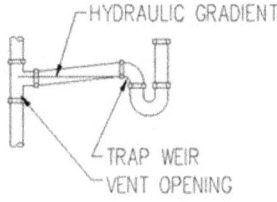

UNSATISFACTORY INSTALLATION

Section 890.ILLUSTRATION (N) Trap Weir/Hydraulic Gradient

Drainage illustration

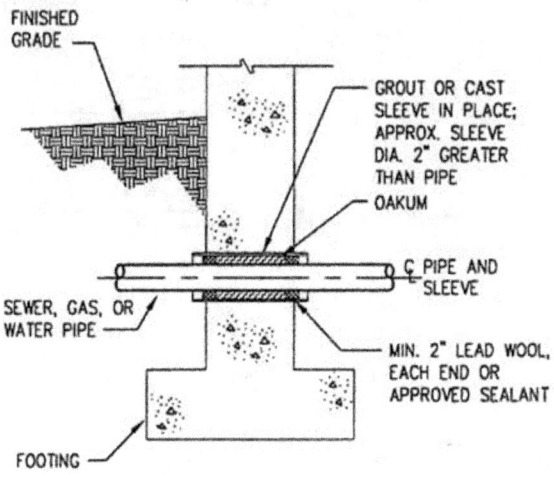

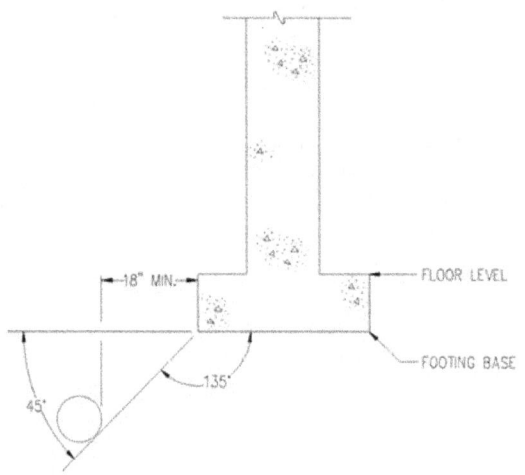

Typical piping parallel to a footing

Vent System

The Venting Theory was proven by connecting a vent pipe to the drain at the trap outlet. The air pressure was kept the same. This method prevented objectionable odors and sewer gases from escaping at fixture waste outlets. (1874)

A properly designed and installed venting system, in conjunction with the soil or waste system is essential to protect trap seals and prevent siphonage, aspiration, or back pressure. The venting system shall be designed and installed to permit the admission or emission of air so that under normal and intended use the seal of any fixture trap shall never be subjected to a pneumatic pressure differential of more than a 1-inch water column. All fixture traps shall be protected by the use of a vent or venting system. In order to allow water to flow smoothly without gurgling, there must be an air passageway behind the water. Vent pipes extend from the drainpipes up through the roof to provide that passage. Vent pipes also carry odors out of the house. In other words, the vent system equalizes pressure in the system.

The drainpipe for each plumbing fixture must be connected to a vent that supplies the pipe with air from the outside. In some cases, the drainpipe is connected directly to a main or secondary stack pipe, which travels straight up through the roof. Plumbing codes strictly prescribe where vent pipes can connect to the stack and how far they should travel. Each building in which plumbing is installed shall have at least one main vent no smaller than 3 inches for each building drain installed. (See Appendix A. Table K and Appendix K. Illustration C.)

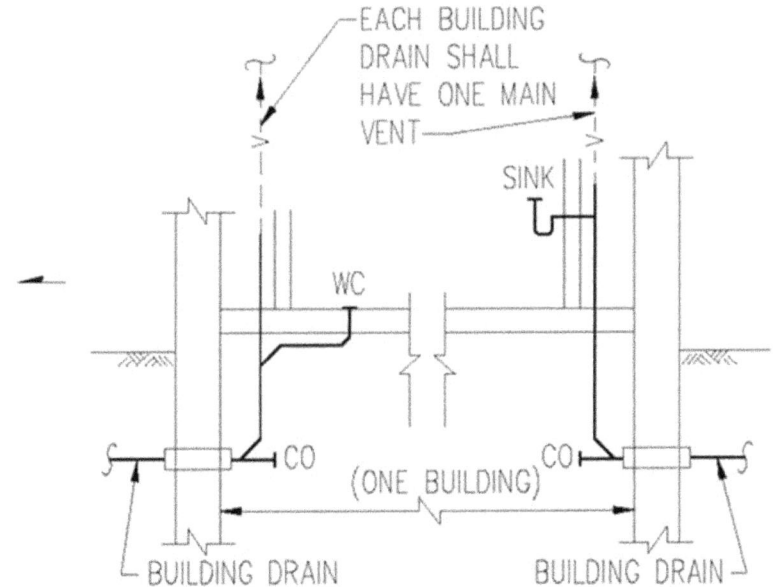

Illustration C

(Referenced in Section 890.120, Definition of "Individual Dry Vent.")

Section 890. Table K Size and length of vents

Size of Soil or Waste Stack	Fixture Units Connected	Diameter of Vent Required (Inches)								
		1¼	1½	2	2½	3	4	5	6	8
		Maximum Length of Vent (Feet)								
1¼	2	30								
1½	8	50	150							
1½	10	30	100							
2	12	30	75	200						
2	20	26	50	150						
2½	42		30	100	300					
3	10		30	100	200	600				
3	30			60	200	500				
3	60			50	60	400				
4	100			35	100	260	1,000			
4	200			30	90	250	900			
4	500			20	70	180	700			
5	200				35	80	350	1,000		
5	500				30	70	300	900		
5	1,100				20	50	200	700		
6	350				25	50	200	400	1,300	
6	620				15	30	125	300	1,100	
6	960					24	100	250	1,000	
6	1,900					20	70	200	700	
8	600						50	150	500	1,300
8	1,400						40	100	400	1,200
8	2,200						30	80	350	1,100
9	3,600						25	60	250	800
10	1,000							75	125	1,000
10	2,500							50	100	500
10	3,800							30	80	350
10	5,600							25	60	250

Agency Note: Per Section 890.1580I, no more than 20 percent of the maximum developed length may be installed in the horizontal position. Vent piping serving floor drains shall be installed in such a manner as to minimize horizontal vent distances.

Section 890.1430 of the Illinois Codebook, Vent Terminals

Vents shall terminate independently above the roof to the outside atmosphere or shall be connected to another vent at least 6 inches above the flood-level rim of the highest fixture served by the vent. Roof Extensions. Extensions of vent pipes through a roof shall be terminated at least 12 inches above the roof unless a roof is to be used for any purpose other than weather protection. If a roof is to be used for any purpose other than weather protection, the vent shall be extended at least 7 feet above the roof.
Vent Terminal Size. Each vent extension through the roof shall be a minimum of 3 inches in diameter and no smaller than the vent that it terminates. Vent terminals shall not be screened.

Section 890.1450 Vent Grades and Connections

All vent and branch vent pipes shall be installed so that they may drain back to the soil or waste pipe. Where vent pipes connect to a horizontal soil or waste pipe, the vent shall be taken off above the centerline of the soil or waste pipe, and the vent pipe shall rise vertically, or at an angle not more than 45 degrees from the vertical before offsetting horizontally or before connecting to the branch vent. Exception: Wet vent and floor drain vents may connect horizontally. (Illustration F)

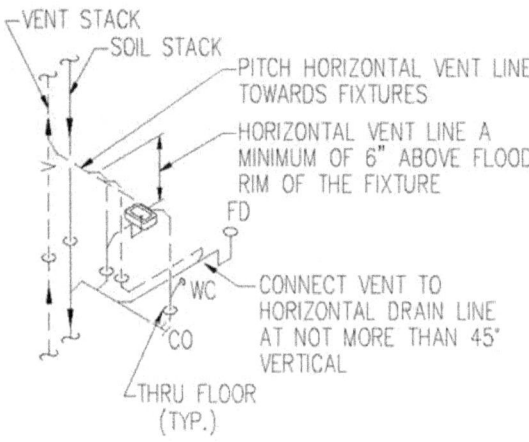

Illustration F

The connection between a vent pipe and a vent stack or stack vent shall be made at least 6 inches above the flood-level rim of the highest fixture served by the vent. Horizontal vent pipes forming branch vents or relief vents shall be at least 6 inches above the flood-level rim of the highest fixture served. (Illustration H)

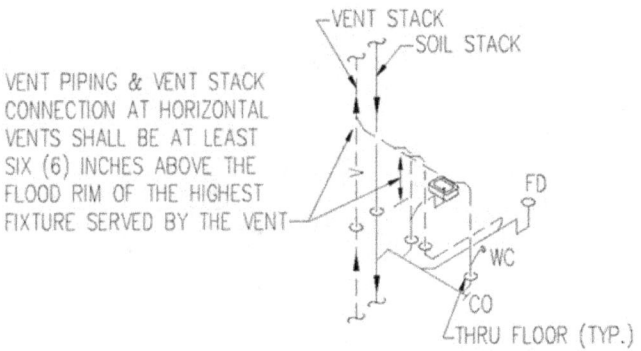

Illustration H

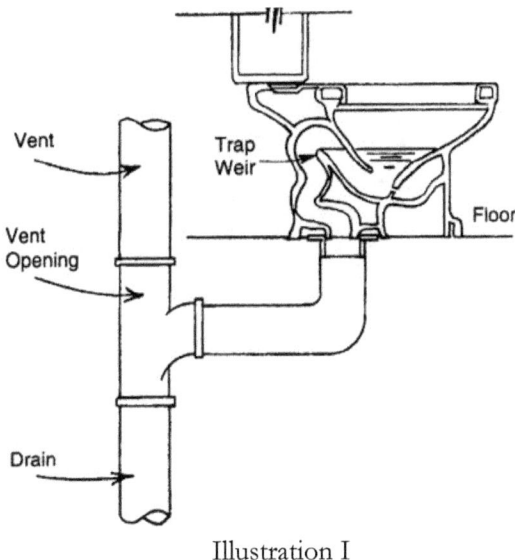

Illustration I

Unless prohibited by structural conditions, each vent shall rise vertically to a point not less than (6) six inches above the flood-level rim of the fixture served before offsetting horizontally. Whenever two or more vent pipe converge, each such vent shall rise to a point at least (6) inches in height above the flood-level of the plumbing fixture, before connection to any other vent. (Illustration I)

Developed lengths

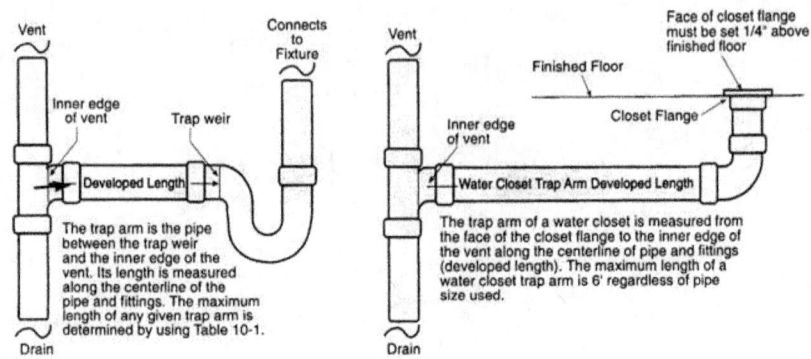

TOTAL DEVELOPED LENGTH
OF TRAP ARM MEASURED
ALONG $\mathcal{C}_L$ = "A" + "B"

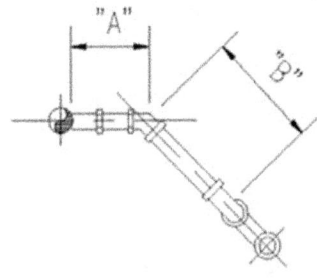

PLAN

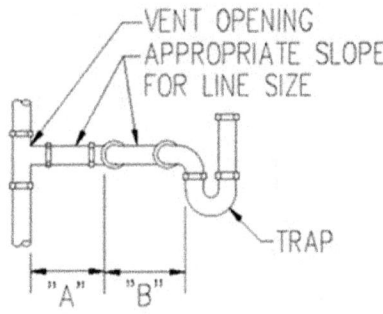

TOTAL DEVELOPED LENGTH
OF TRAP ARM MEASURED
ALONG $\mathcal{C}_L$ = "A" + "B"

ELEVATION

Section 890.1500 Installation of wet venting

Wet vent by definition is a waste pipe that also serves as a vent pipe. Wet venting is most common in conjunction with toilets and sinks; the drain for the sink is also the vent for the toilet. (Illustration Y)

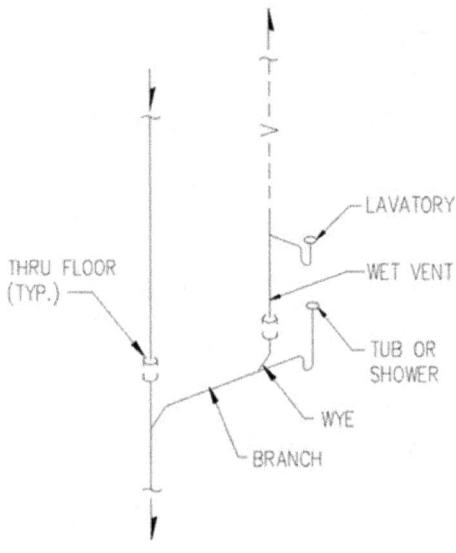

Illustration Y

    a)   The following shall not be used to wet vent another fixture trap: water closets, washing machine connections, blowout urinals, or dishwashers.

    b)   Two lavatories with 1¼-inch traps shall be considered a single fixture for the purpose of this Section.

    3.   A vertical wet vent may be used for two fixtures set on the same floor level, but connecting at different levels in the stack, if the vertical wet vent/drain between the two traps is one pipe diameter larger than the upper fixture trap and that both drains conform to Appendix A. Table I. (Illustration P.)

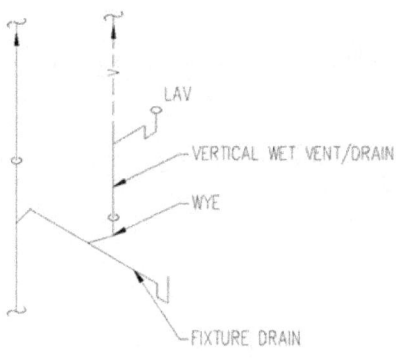

Illustration P

Section 890.TABLE I   Allowed Distance from Fixture Trap to Vent

Size of Fixture Drains (Inches)	Maximum Allowed Distance from Trap to Vent
1¼	2 ft. 6 in.
1½	3 ft. 6 in.
2	4 ft. 0 in.
3	5 ft. 0 in.
4 and larger	6 ft. 0 in.

4. A horizontal wet vent may be used for two fixtures set on the same floor level with one fixture connecting upstream of the other fixture on the horizontal line, provided that the horizontal wet vent/drain between the two fixtures is one pipe diameter larger that the upstream fixture trap. The vent connection shall be located between the traps, and each trap-to-vent distance shall be in accordance with Appendix A. Table I. (Illustration Q.)

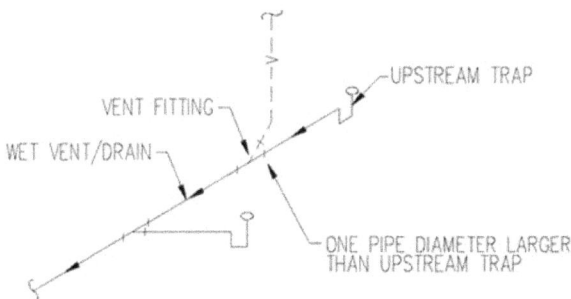

Illustration Q

e) A vertical/horizontal wet vent may be used for two fixtures set on the same floor level. With one fixture, connecting to the vertical stack and one fixture connecting to the horizontal line, if the wet vent/drain is one pipe diameter larger than the upper fixture trap and the drains conform to Appendix A. Table I. (Illustration R.)

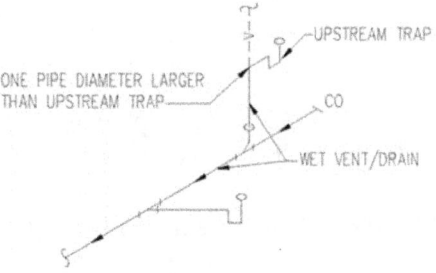

Illustration R

f) A single bathroom group of fixtures, consisting of a water closet, lavatory and a bathtub, shower or floor drain, may be installed with the drain from a lavatory serving as a wet vent for a bathtub, shower or floor drain and for the water closet, provided that:

5. Not more than four DFUs drain into a 2-inch diameter wet vent; and

2) The horizontal branch is a minimum of 2 inches and connects to the stack at the same level as the water closet drain. The horizontal branch may also connect to the water closet bend. (Illustration S.)

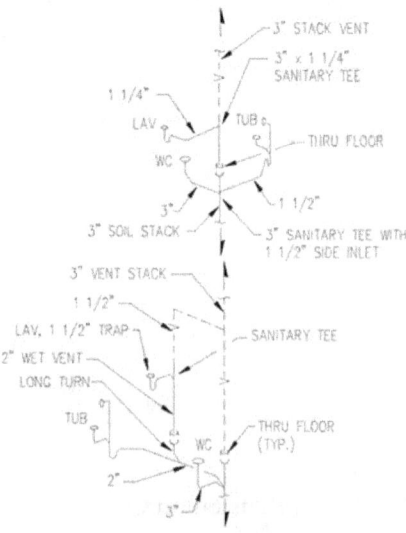

Illustration S

g)   Bathroom groups installed back-to-back consisting of two water closets, two lavatories, and two bathtubs, showers or floor drains may be installed without individual vents, provided that:

   6.   The water closets are wasted to a proper vertical drainage fitting

2)   The bathtubs, showers, or floor drains connect to the stack at the same level as the water closets
3)   The lavatories connect to the stack at the same level

4)   The vent is a minimum of 2 inches in diameter. (Illustration T.)

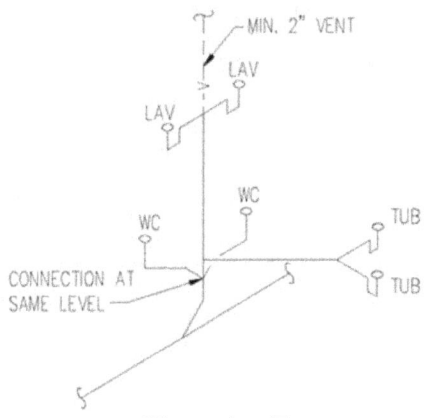

Illustration T

7.   When bathroom groups are connected to the same soil stack, the waste pipe from one or two lavatories may be used as a wet vent for one or two bathtubs or showers, provided that:

8.   The wet vent and its corresponding extension to the vent stack are 2 inches in diameter

2)   Each water closet is provided with an individual dry vent or vertical common vent; and

3)   The vent stack is sized as given in Appendix A. Table
J.   (Illustrations U and V.)

Section 890.TABLE J   Size of Vent Stacks

Number of Bathroom Groups	Diameter of Vent Stacks (Inches)
1 or 2	2
3 to 9	3
10 to 16	4

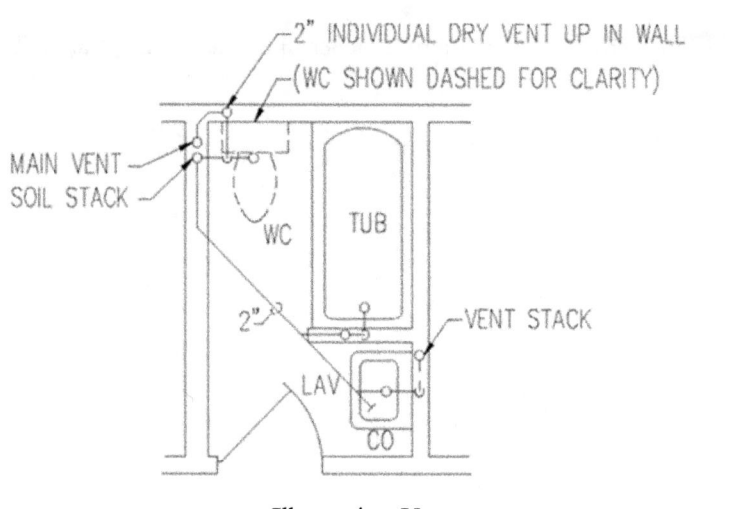

Illustration U

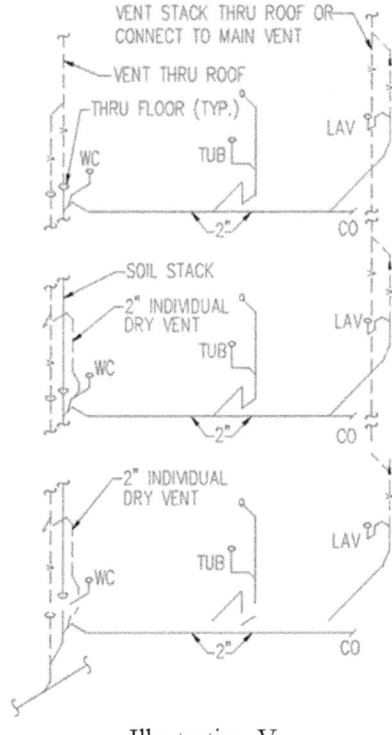

Illustration V

9. On the lower floors of a multi-story building, a water closet may be wet vented with a single lavatory in lieu of the requirements of Section 890.1470(h).

j) Bathroom groups consisting of a water closet, lavatory, and bathtub or shower, connected to a stack by a separate branch, may wet vent the water closet and bathtub or shower with the lavatory, provided that:

10. The water closet and bathtub/shower connect to the stack at the same level:

2) The wet vent and its corresponding extension are a minimum of 2 inches in diameter; and
3) A vent stack connects at or below the lowest fixture connection and is installed for a stack of this type. (Illustration W.)

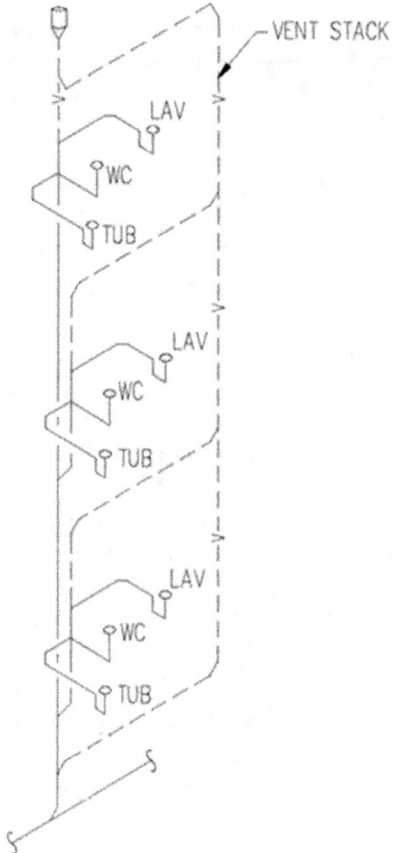

Illustration W

Special Venting

Traps for island sinks and similar equipment shall be roughed in above the floor and shall be vented by extending the vent as high as possible, but at least the drain board height, and then returning it downward and connecting it to the horizontal sink drain immediately downstream from the vertical fixture drain. The returned vent shall be connected to the horizontal drain through a sanitary drainage fitting. In addition, shall be provided with a vent taken off the vertical fixture vent by means of a sanitary drainage fitting immediately below the floor. In addition, extending to the nearest partition and then through the roof to the outside atmosphere, or may be connected to other vents at a point at least 6 inches above the flood level rim of the fixture served. Drainage fittings shall be used on all parts of the vent below the floor level, and a minimum grade of ¼ inch per foot back to the drain shall be maintained. The returned bend used under the drain board shall be a one-piece fitting or assembly of a 45 degree, a 90 degree, and a 45-degree elbow in the order named.

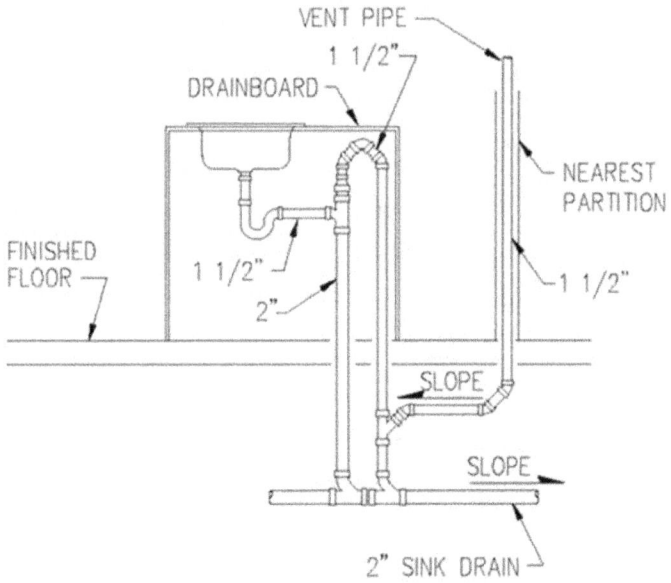

Illustration GG

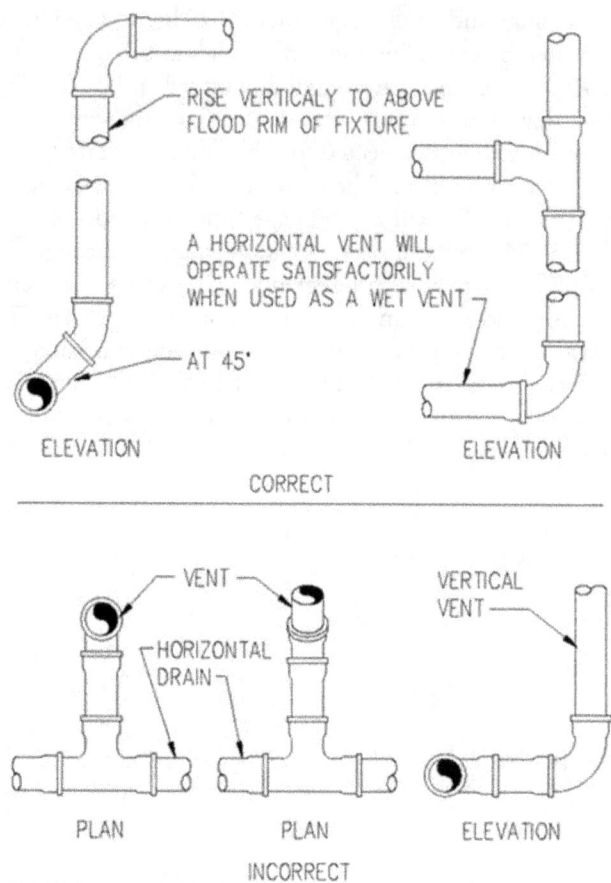

Illustration AA
Vent connections

Section 890.1570 Vent Headers

Connection of Vents. Stack vents and vent stacks may be connected into a common vent header at the top of the stacks and then be extended through the roof to the outside atmosphere at one point.
This header shall be sized as provided in Appendix A, Table K. The number of units being the sum of all units on all stacks connected thereto, and the developed length being the longest vent length from the interception at the base of the most distant stack to the vent terminal to the outside atmosphere, as a direct extension of one (1) stack.

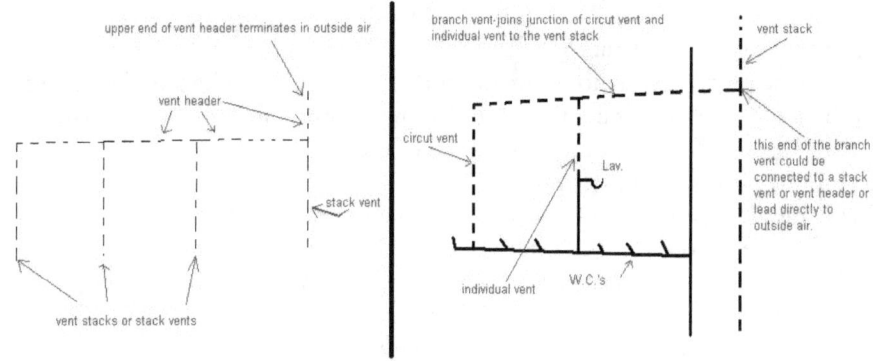

13 ESTIMATING PLUMBING WORK

Medium to Large Projects

Estimating can be a challenge even for some licensed plumbers. Most apprentice plumbers do not estimate work at all until their third or fourth year in the trade. With that said, there are also many apprentices giving estimates on a daily basis. So many plumbing companies today have cut out hourly pricing altogether and have implemented a flat rate for services.

Many plumbers are left in the dark as to how to properly "bid" a job. I have had the pleasure to work on both sides of this fence. I am very familiar with the days of figuring work on a time and material basis, as well as a flat rate.

Plumbers' rates vary significantly by location. In some areas, plumbers charge $45 -$75 an hour; in other regions, the hourly rate can be $75 – $150. Most plumbers charge a two-hour minimum or a service call fee of $75 - $150, and some plumbers bill a flat fee per job instead of an hourly rate. Either away, exact costs will depend on the complexity and type of work done. For example, it can cost $50 -$200 to have a plumber repair a toilet by replacing the tank mechanism, wax ring, or flange. Replacing a toilet typically costs $150-$600 for standard installation of a basic toilet.

The following excerpt is the same one I have introduced apprentices who have taken my class, and this, by the way, is based on a larger scale-plumbing project with blueprints. Later in this chapter, I will show you a few fast tract ways to put a smaller project bid together quickly and effectively.

Typically you would be called upon to bid a plumbing project, there may or may not be a job site meeting. If this is an option, I highly recommend attending. You can learn many details that a blueprint may not provide. Such as meeting the general contractor, your competition and so on.

Often the scope of work may change slightly from the drawing. This is also a perfect opportunity to find out about draws, or payments.

When in attendance at a job site meeting there are a few things to consider bringing with, a quality LED flashlight, a note pad, and a tape measure are among the top three. If plans have been sent electronically, it is a great idea to bring a copy with. You can take the necessary notes directly on the copy of the plans.

In most cases, the invitation has come directly from a general contractor; this is a great time to "size" him up as well. Be leery of a fast-talking, disorganized, or unscrupulous contractor. If he is looking to cut corners, or he tells you "bid this one low and I'll have more work for you," That is a red flag! You should not bid a job like this.

If given the chance, it is a good idea to do a quick internet search on the contracting firm. You can find reviews, how long has the company been in business, what some of the projects they have been awarded are, and so on.

Let us assume the job site meeting is a success, and you now have three days to figure an estimate for the project. Your head maybe spinning at this point with many questions running through your mind, such as do I have the labor for this project? What will my number look like compared to my competition?

First, if work force has come in to question this is probably not the job for you. If the plans have been sent to you via email, I recommend downloading them to a flash drive. Form there you can take the flash drive to an office supply store and have them printed into a set of easy to read drawings.

Once the plans have been flipped through for the first time, the one page that will require a bit of reading is the plumbing notes. This page of the drawing will have specifics pertaining to the scope of work. This will include such things as choice of materials used, what codes are to be followed, and other details.

The plumbing notes should be reviewed with care. The next thing to look at closely are the details as to what plumbing fixtures have been selected. This will include the specifics regarding the hot water tank and any other equipment.

This is a good time to start a list or a take-off of the fixtures listed. Using your take-off list include everything you will need to completely install the given fixture. In other words angle stops, supply line, wax rings and so on.

Once you have completed the fixture take-off, you can fax this or email it to your plumbing supply house sales representative. This is one of the first things to be done for two reasons. One, this list can be compiled in a short amount of time. Secondly, while waiting for the supply house to get your cost of the fixtures, you can immediately begin to work up a cost for the man-hours to install the fixtures.

Plumbing Fixture Quantity Take-off Worksheet								
Project:				Date:				
Clear Values from previous calculations				Page ___ of ___				
Section:				Takeoff by				
Fixture Type	Floor Level							Total
	Basement	1st	2nd	3rd	4th	5th	Roof	
Total								

An example of a fixture worksheet can be created on either a word document or excel sheet in short time. Hopefully, by the time your fixture cost is sent back, you would have finished the labor or man-hour cost associated with the fixture install.

There are many types of software programs to assist in estimating large jobs, however, they can be quite costly which is something to consider if you are a new company. Generally, it is a good idea to highlight the drawings; this is a very important step because it is really the point at which you get familiar with the job. Here is an example of what and how you could highlight, of course the colors chosen can be completely up to you.

Water Piping – Red (quite a few contractors that highlight hot water in red and cold water in blue which is fine, however it's all water piping and it makes no difference on an estimate.)

Fixtures and Mechanical Equipment – Orange (use orange for all plumbing mechanical equipment i.e., floor drains, clean-outs, floor sinks, lavatories, toilets, urinals, kitchen sinks, water heaters, recirculating pumps, backflow device's, and house pumps. Consider these a separate step in the plumbing installation process, its okay to lump them together. You can choose to separate plumbing equipment and plumbing fixtures.)

Waste & Vent Piping – Green
Underground Waste and Vent Piping – Yellow Drain Tile – Brown

After the take-off is complete, it is a good practice to compare the plumbing drawings to the architectural drawing. There are times when the specifications for plumbing fixtures do not appear on the plumbing drawings but do appear on the architectural drawings. Architectural drawings take precedence over all trade drawings, then the architectural detail drawings are next, then the trade drawings are last.

It is a good idea to have the mechanical drawings printed for review, because the mechanical drawings have cooling units on some of the floors especially for data centers. These units have make-up water connected to them and backflow devices will usually need to be installed.

It is also perfectly acceptable when writing the proposal, to exclude any work you will not do as a part of the scope.

Another consideration is whether you would have need a sub-contractor. For example, if the drawing indicates there is a concrete floor to be saw cut open, in order to demo existing plumbing. The plumbing notes will tell you (possibly) that the concrete demo is a portion of the plumbing scope and is to be included with the plumbing bid.

Sub-contractors may also be needed for other considerations that may be indirectly related to your work as the plumbing contractor.

A few reasons to hire a sub-contractor beside the concrete sub might also include a pipe insulator. Most construction projects require water pipe to be insulated; also, the 2015 energy code requires all hot water piping to be insulated. Fire stopping could be another requirement, often when plumbing pipes penetrate through a parking garage fire stopping is required. If you are going to be installing new sewer and water service, will you need a sub-contractor?

These are just a few examples of things to consider, and a worksheet will help.

Phase: General or Special Conditions

DESCRIPTION	QUANTITY	UNIT	TOTAL ESTIMATED MAT. COST	UNIT MAN DAYS	TOTAL ESTIMATED MAN DAYS
Temporary Pumps, Piping, etc (See Recap)					
Temporary Toilets (See Recap)					
Shanty/Office Rental					
Breakage Allowance					
Saftey Act Costs					
Supervision (Non-Productive)					
Hoisting Charges					
Mason Hydrant					
Pipe Rack & Pipe Shop					
Telephone					
Intercom System					
Sleeving (Except Typ. Floors)					
Inserting (Except Typ. Floors)					
Shop Drawings (Submittal Procedures)					
Engineering					
As-Built Drawings					
Blueprinting and Sepias					
Packing of Sleeves					
Expendable Tools					
Clean Debris					
Tag Valves & Schedule					
Equipment Instructions					
Cutting & Patching					
Scaffolding (Rental Only)					
Watchman					
Welding Consumables					
Travel Time					
Carfare					
Parking Fees					
Drayage or Trucking					
Incidentals					
Job Site Truck					
Pro Rata Charges by G.C. (Or Architect)					
Core Drilling					
Pipe Labeling					
TOTAL THIS SHEET					

FORM #G-1

Job Name: _____ Estimate No.: _____ Date: _____

Phase: Plumbing

DESCRIPTION	TOTAL EST. MATERIAL COST	QUANTITY	UNIT	COST/ DAY	TOTAL EST. LABOR COST
1. Total General & Special Conditions			DAYS		
2. Total Fixtures			DAYS		
3. Total Mechanical Equipment			DAYS		
4. Demolition			DAYS		
5. Underground Piping Material			DAYS		
6. Rough-In Typical Floors Material			DAYS		
7. Waste & Vent (non-typ. floors) Material			DAYS		
8. Water Piping & Nippling (typ. floors) Material			DAYS		
9. Water Piping (non-typ. floors) Material			DAYS		
Add 3% to total of 5-6-7-8-9			DAYS		
Sub-Total					
Deduct Buying					
Premarkup Adjustments "R" Sheets					
Escalation (See Recap)					
Material & Labor Sub-Total					
Sales Tax					
Sub-Total					
Review Adjustments					
Sub-Total			Material Carryover		
TOTAL LABOR & MATERIAL					
Overhead					
Sub-Total					
Profit					
H. Permits/Bonds/Fees					
J. Subcontractors					
Markup on Subcontractors					
Subcontractor Carryover Sub-Total			Subs Carryover		
Sub-Total					
Performance Bond (If Required)					
TOTAL THIS SHEET					

FORM #E-1

Phase: Subcontractors

DESCRIPTION	QUANTITY	UNIT	UNIT PRICE	TOTAL ESTIMATED MAT. COST	UNIT PRICE	TOTAL ESTIMATED LABOR COST
Pipe Covering (Insulation)						
Backflow Certification						
Electric Tracing						
Fire Safing						
Power Wiring						
Control Wiring						
Trench Box Rental						
Water Tap Contractor						
Exterior Utilities						
Sewers & Excavating (Operator & Machine)						
Haul Surplus Dirt						
Meter Vaults						
Super Grease Basins						
Chlorination						
Core Drilling						
Saw Cutting						
Warehouse Space Rental						
Chemical Toilets						
Painting						
Special Hoisting of Material/Crane Rental						
Millright						
Equipment Hoisting						
Metal Drip Pans						
Bathtub Patching						
Hi Jacker Rental						
Sprinkler Work						
Pump Rentals						
Rubbish Removal (Scavenger)						
Wellpoints						
Soil/Proctor Tests						
Rentals Special Tools or Machinery						
Tub/Shower Enclosures (If Subcontractor)						
Scaffold Rentals						
Vanity Tops (Only If Subcontractor)						
Vanity Cabinets (Only If Subcontractor)						
Toilet Partitions						
Toilet Room Accessories (Only If Subcontractor)						

FORM #S-1

General or special conditions are real or potential cost you should consider for every project. They are also a checklist of items that you may want to qualify out of your proposal. That is why it is important to read all the notes on the architectural, plumbing, and mechanical drawings and also the General Conditions and Mechanical Systems of the specifications.

Temporary Pumps, Piping – Sometimes a plumber needs to install temporary piping to provide water for the rest of the trades or do temporary sump pumps or domestic water booster systems.

Temporary Toilets – Sometimes during a rehab or a complete gut of a building, the plumbing contractor will activate or build temporary toilet facilities for the project. Office Rental – I have been on many large projects where a temporary office or trailer is needed to house supervision and workers. You can rent temporary office trailers or storage trailer in order to house supplies and material. The last step in preparing a bid is to tie all of your worksheets together in a final or overview sheet to use as a checker to be certain nothing has been left out. Now I have left out several other things such as warranty or punch-list line items, however this would be for use on a medium to large project.

You could however think about things such as travel time, parking, additional non-productive labor that of a supervisor, rental equipment, and so on. This list could become extensive depending on the size of the project. I think you get the idea.

Remember these are not set in stone but they are a starting point. Here are the phases and a brief explanation of each.

Phase 1: Total General and Special Conditions – Transfer your totals from the General and Special Conditions page to this page and extend.

Phase 2: Total Fixtures – You may want to combine Fixtures and Mechanical Equipment however if the job is massive, you can separate. Separating the two on a big project helps to put together the schedule of values on your billing paperwork. Here is what to consider plumbing fixtures, toilets, lavatories, urinals, bidets, kitchen sinks, bathtubs, showers, mop basins, or slop sinks, drinking fountains or water coolers, hand sinks and emergency eyewashes. All plumbing fixtures and their related faucets, P-traps, supplies, strainers, and install hardware.

Phase 3 : Mechanical Equipment – for the sake of this piece mechanical equipment are floor drains, roof drains, floor cleanouts, wall cleanouts, floor sinks, backflow devices, sump pumps, ejector pumps, house pumps, water heaters, boilers, re-circulating pumps, mixing valves, grease separators, triple oil basins, fixture carriers, flashings, etc..

Phase 4: Demolition – this encompasses only plumbing demolition.

Phase 5: Underground piping – this is all piping below grade.

Phase 6: Rough-In Typical Floors – This is your waste and vent material in a typical floor situation. A typical floor is a floor in a multiple floor building where the plumbing is the same on each floor. It is much easier to put an estimate together if you have typical floors. On the material side, you take off one floor and multiply it by the number of typical floors in the building. About labor, you do have to consider what can be called the "experience factor". The "experience factor" is the experience your labor force gets by doing something repeatedly. If you start setting toilets, you are much better and faster on the last one than you were on the first.

It is the same with rough in, your plumbers are much better on the fifth floor rough in than they were on the first. At least they should be better; your estimate can depend on it.

Phase 7: Waste & Vent Non-Typical Floors– You will probably use this phase more than the one above. It is your waste, vent material, and labor for non-typical plumbing.

Phase 8: Water piping Typical Floors

Phase 9: Water piping non-Typical Floors

Fixture Bidding

How to fast track a smaller plumbing project, Mr. Jones called and would like to have a bathroom installed in his basement. Here is an easy way to figure out what to propose as your bid. First thing to do is ask questions be certain to ask about fixture selection and who will provide them. There is going to be a 99.9 percent chance you will also be cutting up concrete, hauling spoils away and possibly patching the concrete back.

Do not hit the panic button this can be figured out using a "per opening" price. Much the same as with flat rate pricing. For an example if the project is within Chicago we already, know that the materials used for underground are going to have to be cast iron. If the project is outside of Chicago, we can utilize PVC. We will price both jobs separately; first, we will take the project in Chicago. As I pointed out, we know our choice of materials for our underground will have a much higher cost than that of PVC. With that in consideration, we typically will charge up to $2,850 dollars "per- opening." Now let us say the bathroom will consist of one water closet, one lavatory, and one shower base. We will also supply or furnish all fixtures, necessary permits and concrete removal and patching.

Simply put 2850 times three is $8,550 now we need to add $100 dollars for a permit and $35 dollars per square foot of concrete work. A typical bathroom is going to require a five-foot-by-five-foot area of concrete removal. Five times five is 25 or 25 square feet of concrete work. This means we need to add $875 dollars to allow for concrete.

What we now need is to total these numbers to give us $9,525 for our estimated bathroom install. If we break this down the bathroom addition for Mr. Jones, residing in the city of Chicago would be a total job including fixtures, waste, water, and vent complete turnkey.

At a price of $9,525.00 dollars, of course, walls paint and finishes would not be in our plumbing bid. If the same exact project is located in the suburbs, again we know per state plumbing code PVC is acceptable for underground. We can apply the same technique as listed above, however the "per-opening" rate would be less.

Mr. Jones has now moved to the suburbs and would like the same job as when he lived in the city. The only difference is his new home has an overhead sanitary sewer rather than a gravity system. What we know need to consider is we may have a lower rate working with PVC; we have now added an opening (ejector pump) to the project and have a bit more concrete to remove.

Our "flat rate" or opening rate may only be $2,520 but with the additional opening (four total) we are at $10,080 as a base. We still need to add our $875-$925 for the concrete work. As you can see even with a 15 percent decrease on the "per opening" our bathroom in the suburbs will actually be more expensive than our city project.

Both of these are just an example of how to price out a project while on site. Keep in mind the "per opening" rate as in the example is just that. You will have to do your own homework to come up with an all- inclusive rate.

Just as important as figuring the job, it is equally important to "pitch" the job or present your price to the potential buyer. This can be accomplished in to ways, one a formal typed proposal of the work, or two a verbal presentation. Typically, for work in excess of two thousand dollars it is a good idea to have a written proposal. Consumers prefer a complete breakdown or line item of exactly what you propose to do for them. In addition, a formal written proposal is something tangible that a customer can have in hand while making the decision of who to hire.

A typed proposal should include as much detail as possible with as little plumber jargon as possible. Let us assume the job is a clean out install, be sure to add all the steps that it will take to achieve the goal. What I am referring to is do not just write down dig front yard and install clean out.

Include details such as; excavate sanitary sewer five feet from outside foundation to a depth of six feet. Install one six-inch multi directional tee, bed pipe with CA-7 granular fill; connect to existing six-inch clay pipe with no-shear couplings. (Company name) will obtain necessary permits and arrange inspections. (Company name) will haul away all spoils and debris from site.

It is always a good practice to spell out what your price will include, try to stay away from what your price does not include. Remember your customer is expecting to get the most for their dollar as possible. The last thing they want to hear is this job is going to cost $5,000.00 dollars and here is everything I am not going to include.

You want to "sell" as much value as possible, the proposal should take a little time to complete, and whenever possible provide at least two options or as many as three. When typing a proposal be, clear as to what each scope of work will provide.

During the verbal presentation, again be clear as to what you will provide. Be confident, friendly, and polite. A customer does not want to hear, "wow! That is a mess, you should really think about getting the whole basement re-piped." You are the professional that should have solutions not snide remarks.

14 CUSTOMER SERVICE

Plumbing Hats

Plumbers wear many different hats, not just trade related as electrical, carpentry, drywall, and sometimes-even therapist. There is I'm sure more than a few of you reading this thinking why is there a chapter about customer service in a plumbing book? Simply put, if your business is going to be successful it starts with great customer service.

The difference between good and great is often determined by what you do to push yourself beyond the comfort zone. This is true on the technical side of plumbing as well as great customer service. Plumbers wear many different hats. All too often I have sat across the interview table and have heard "I'm the best plumber" well, that is great but in the service business that will only get you 40 percent of the way there.

The other 60 percent of plumbing service is how well you are with your customer. I have also had veteran plumber's interview and tell me "I'm not a salesman." I would like to clear that myth as well; we are all sales representatives in one way or another. A service plumber must be patient, courteous and a good listener. After all, the customer deserves your undivided attention.

Think about some of the positive customer service experiences you have had. Did the person you were talking to make you feel like your problem was important, that your business mattered, and that they wanted to help you anyway that they could?

When a customer calls for a service, and this could be anything in the service industry, primarily the service call is and should be all about them not you. Remember, when a plumber shows up at someone's home it is usually at his or her worst moment; consider this when knocking on the door. It is up to you at that point to put on the "compassion hat."

Have the "compassion hat" on at the door, have a homeowner's mindset when first greeting the customer. Remember, water is either leaking, or backing up. The average homeowner did not have the four years of apprenticeship, attend any type of plumbing training, or has received bad advice from their brother-in-law who used to be in the construction industry.

One thing that the customer knows without question is they have an issue and want it repaired for as little money as possible. This is the time to put on the "confidence hat." Re-assure the customer through confidence that you are the person that can fix the issue and make things right again.

Confidence is the key word, there is a line between confidence and overzealous, or cockiness. Know the difference and at all cost avoid the later. The best way to demonstrate value to a customer is to provide solid troubleshooting skills without the arrogance. I have a few basic rules that I explain to a new apprentice when it comes time to demonstrate value. Of course, it is time to put on the "value hat."

Number one, if you are the second or even third plumbing contractor on the job, never under any circumstances bad mouth the last contractor. This would defiantly support arrogance. Not only that, you could be insulting the customer, or another plumber within your company.

Number two, always listen, ask questions, and be engaged with the customer during you analysis or "investigation." Troubleshooting is a skill that can only be learned with time and patience. Do not overthink a situation when it comes time to evaluate a customer's issue.

Sometimes, the hardest thing for a plumber to do is to just listen to the customer. Being in a hurry to close a call and not taking the time to listen can prevent the customer from feeling that they can trust your company.

Number 3, clear communication with a customer is essential. As I stated earlier in the book avoid using plumber jargon when speaking to a customer. You will have them lost within seconds. Again, they did not receive the training that you have. Understand that they are feeling vulnerable, if you try to explain the problem with plumber jargon the customer may become frustrated even more.

Number four, there is an old saying "you'll catch more flies with honey" is so true. Customers will respond in a positive way if you are polite. Leave your bad day out in the truck before knocking on the door.

If you are able to show the "value" to your customer, and remember even though you may represent a large plumbing company, they are at that moment in time YOUR customer. Your fees may not be frowned upon as much as the plumber that shows up and is arrogant.

Number five, all too often I have witnessed apprentice's "shy" away from informing a homeowner exactly what is wrong with their plumbing. In addition, will only correct a portion of the issue. I will give you an example of a drain call, where an apprentice can visually see the piping has begun to deteriorate.

If the apprentice would, again, switch hats and inform the customer of the deteriorating pipe and offer a solution, rather than just clear the drain. What will end up happening is when the drain clogs again, and another plumber shows up. He will immediately point out the issue, the first thing a customer is going to say "is why the other guy didn't say something?" In that scenario, the first plumber did not demonstrate confidence at all. Again the homeowner did not have any training in plumbing, if they did chances are your company would have never received a call.

Good economy or bad economy, people always want value. Value is defined as the ratio of usefulness to cost— what you received compared to what you spent. To determine what a customer values one must learn how to really listen to the customer's communications.

This is more than what is printed in specifications; it comes from discussions in face-to-face meetings. It comes from asking many questions and "listening and understanding," not to rebuttal or sell.

A challenge for most contractors is to really listen to the voice of their customers. Few have any formal systems to hear, collect, analyze and respond to the voice of the customer.

Is The Customer Always Right?

Research, has shown that responding to a complaint quickly and fairly will enhance customer loyalty. The opposite is also true. A formal complaint system does not require purchasing expensive software or writing volumes of standard operating procedures. It requires establishing a process for how any complaint, which comes to anyone in the company, is addressed.

The process answers these key questions: Who owns the complaint, what action was taken, who keeps the customer informed of the handling of the complaint, how do we know the complaint was resolved, and how do we know the complaint's resolution was acceptable to the customer.

Besides addressing the complaint, the company needs a systematic way to determine if the complaint is a random occurrence or a trend. Have similar complaints been received? How often? Acting quickly and fairly to resolve complaints can help keep loyal customers, but customers do not really want to make complaints in the first place.

It adds to the cost part of the value equation. Prevention should always be the higher focus. Spotting trends in complaints can alert managers to seek the root cause and implement countermeasures.

In short, depending on the circumstance, the customer is always right. One thing to consider is if you upset your customer, they will leave negative reviews online. If you are a smaller company, one negative review can be devastating. Best practice is to eliminate complaints quickly. I have taken complaints in the past and actually turned a complaint into an up-sell. (Be compassionate). Customers do 100% of the buying. They also do 100% of the repeat buying. Moreover, they do about 100% of the referring. Previous customers spend their money 33% faster than first – timers. They are 40% more likely to buy the up – sell offer. They are three times more likely to refer others. In addition, they cost one-sixth the amount to retain as they did to attract initially. Think that through for a moment. You spend roughly $275 to $325 to get a customer. You get nothing to watch them leave in addition, when they leave, here could be why:

74 percent blame poor customer service as a major factor in their decision, which also included the following.

55 percent say "company indifference" or "no contact." More than half leave you because you showed no interest in them staying!

32 percent are disappointed with the quality of work.

25 percent feel pricing was unfair. Many contractors think this is the No. 1 reason, but actually, they were not given enough "value" reasons to stay.

9 percent blame functionality. Their equipment did not work.

8 percent say their needs changed. They felt they did not need your business anymore.

The lack of a Customer Retention program is the largest mistake you can make for your business. Yet most plumbers fall into that category.
You will never keep a customer if your service is slipshod or poorly thought out. Customer Retention begins with the first interaction a person has with your firm. The key to making those buyers repeat customers is proving that you deserve and value their patronage.
Here are four ways to differentiate yourself through service:

1. Call customers with status reports (even if everything is moving along OK). A 30-second phone call to say your tech is late is invaluable.

11. Extend the relationship beyond the buying and selling. Done largely with follow-up contact such as thank-you calls, cards, and newsletters. Remember these are not "selling" they are building something more valuable: Loyalty.

12. Help your customers know what your company offers and does. Ever had customers say, "I didn't know you did that!" right after they paid your competition to do it? Highly unpleasant learning experience. Newsletters accomplish this, but so do "on hold" messages and broad market blasts.

13. When you blow it (and you will — you are human!), make amends quickly.

Your business will not grow without customers. Your customer base will not grow without service. Solid customer service is the first step toward customer retention. After you have made the first step with your stellar service, the rest is simply building on that initial investment — an investment that will return to you in referrals, repeat purchases and upsells. The best part is that you can do it. It is not for "other" contractors to do and you copy. It is for you to do to be first and different, and it is easy. (Get out of the comfort zone).

Contact them four times a year. At least send them a newsletter, news card, customer survey or thank-you card. You also may send a maintenance agreement letter, preseason, or post-season special, customer-only discount postcard, free or discount service offer or "happy calls" following a service. Help your customers feel more "inside," connected or have some input for complaints, suggestions, questions, etc. It strengthens the relationship. Without a relationship, they leave you.

Want to be a hero? Inform your customer of a price increase that is coming, but you can still do it at the "old" price if they hurry ... Staying in touch through a newsletter helps.

Conduct customer surveys annually. All you are doing is increasing the flow of knowledge and getting *free marketing research*.

Simply ask, "Are your utility bills getting higher? Have you noticed an increase in repairs? Have you heard of a product or service you would like us to add? How could we improve our service to you?" At the end, you offer a discount on the next service for sending it in (or a small gift certificate). Increase the reward if they will put two or more referrals on the bottom. You will not believe how much value this will give you *and* your customers. Remember, educated customers buy and refer more. Your job is to make sure they stay educated about you. The big lesson in all this: Your customer list is the most important single thing in your business. Period. Extracting value from it is the most important business mission. If you lose 10% of your customers a year — the industry average — what did that cost you.

It costs you far more *not* to have a Customer Retention program than to have one. Build the list, build the relationship, and keep the customers.

We have ended up back at the 60/40 split of knowing the difference between a good plumbing provider and a great plumbing provider. Remember, in the service business you should strive to become 60 percent customer care and 40 percent plumbing genius with that as a goal, or mindset you will be successful in this industry. The plumbing portion of this business is the easy part to learn it is the other 60 percent, of this business (customer service) that can be difficult to get to a master level.

15 EXAMINATION STUDY GUIDE

Exam Qualifications

So you are ready to take the Illinois state plumbing exam. Not so fast let us first determine a few things. First, have you met the requirements that the state has mandated? Are you mentally prepared? Have you spent time reading and reviewing the Illinois state plumbing codebook?
There are sections within the Illinois State Plumbing Codebook that will point this out in detail. We will be looking closely at this first that way you will be able to determine if you are qualified to take the exam and, fill out the appropriate application.
The section that I am referring can be found in the Illinois State Plumbing Codebook under Illinois Plumbing License Law. Before we jump in, we first need to establish the definition of an apprentice plumber. Apprentice plumber means any licensed person who is learning and performing plumbing under the supervision of a sponsor, and who is an Illinois licensed plumber.
In short, if you have not been sponsored as an apprentice under a licensed plumber, you do not meet the basic requirements for the exam. The following section are the requirements for admission to the plumbing license exam.

Section 750.300 Requirements for Admission to the Plumbing License Examination

To apply for admittance to the examination for a plumber's license, a person shall file an application for examination on forms provided by the Department.

14. The application form may be obtained by downloading the application from the Department's website (http://dph.illinois.gov/topics-services/environmental-health-protection/plumbing

2) The application shall be submitted to the

Illinois Department of Public Health, 525 West Jefferson Street, 3rd Floor, Springfield, Illinois 62761.
The Department will accept applications postmarked at least 30 days before the examination date. On each examination date, not more than 50 applicants (not more than 40 during winter months) will be examined. The examination will be scheduled at least once every three months. The Department and the Board may schedule additional examination dates as they deem necessary, based on the number of applicants.

4) The Department and the Board will establish examination dates and locations. This information will be included with the examination application form.

15. For each application, the following materials must be received by the Department, postmarked at least 30 days before the examination date:

16. A completed application form;

2) A photograph of the face of the applicant at least 1½ inches by 2½ inches.

3) Proof of eligibility as specified in subsection I; and

3) The required non-refundable application fee as specified in Section 750.1100.

Listed below as noted are the required fee structure for the exam and failure to pass the exam.

Section 750.1100 Plumbers' and Apprentice Plumbers' Examination and Licensure Fees

The applicable fee shall be submitted to the Department with each application for examination, licensure, or certification as follows:

a) Plumber's Examination Fees as of January 2014

1) Plumber's License Examination Fee when applicant is licensed as an Apprentice Plumber in Illinois $175

2) Plumber's License Examination Fee when applicant is registered or licensed in a state other than Illinois $225

3) Plumber's License Re-Examination Fee $175

4) Plumber's License Re-Examination Fee when applicant is registered or licensed in a state other than Illinois $175

Continuing section 750.300

Section 750.300 Requirements for Admission to the Plumbing License Examination

The applicant shall be a citizen of the United States or shall have declared his or her intent to become a citizen. (Notarized papers, such as "Intent to File for Citizenship", shall be submitted to the Department.)

17. The applicant shall have completed at least a two-year course of study in a high school, or an equivalent course of study, equal to 10 credit hours.

e) To be eligible for the plumbing license examination, an applicant shall possess one of the following combinations of experience and education and shall provide proof of experience and education as follows:

18. Illinois licensed apprentice plumber:

19. Each applicant shall have served a minimum of four years as an Illinois licensed apprentice plumber.

B) Each applicant who has served an apprenticeship shall be able to establish that he or she received instruction through practical experience under the supervision of a licensed plumber.

C) The term of apprenticeship shall be not less than 1,400 hours per year, for a total of 5,600 hours in four years.

2) Illinois licensed apprentice plumber with training or education:

20. Each applicant shall have served at least two years as an Illinois licensed apprentice plumber and have two years of approved courses in plumbing (see Section 750.540) for a total of 5,600 hours.

Section noted 750.540 as follows:

Section 750.540 Topics for Approved Programs of Instruction in Plumbing and Approved Continuing Education Courses

Approved courses of instruction in plumbing shall provide instruction in the topics specified in subsections (a) through (dd) below. Approved continuing education courses shall provide instruction in at least one of the topics specified below.

    21. Public Health and its relationship to plumbing

    22. State of Illinois Plumbing License Law

    23. Administration and enforcement

2) Licensing of apprentice plumbers and plumbers

3) Plumbing code requirements

4) Plumbing inspection

    24. Basic principles of plumbing;

    25. Planning and designing a plumbing system including estimating, installation, repair, maintenance, alteration, extension, and dismantling

e) Plumbing materials, fixtures, and equipment

f) Joints and connections;

g) Traps and cleanouts;

    26. Interceptors and separators;

    27. Hangers and supports;

j) Indirect waste piping and special wastes;

k) Water supply and distribution system;

l) Public and private water supply systems;

m) Drainage system;

n) Private sewage disposal systems, municipal or public sewage disposal systems, and/or sanitary districts;

o) Vents and venting systems;

p) Inspection and testing of a plumbing system;

q) Sciences of pneumatics and hydraulics as they apply to plumbing

r) Safety devices allied with a plumbing system

s) Hot water systems and water heaters;

t) Soldering, welding, caulking, and wiping

u) Copper material plumbing system

v) Glass material plumbing system

w) Plastics and thermoplastics material plumbing system;

x) Cast iron plumbing system, including Durham system;

y) Job safety

z) Use and care of tools and equipment;

aa) Handling and disposition of wastes that would damage a plumbing system and sewage disposal facilities;

bb) Alternate plumbing systems;

cc) Solar plumbing systems; and

dd) Pumping of wastes.

B) Proof of practical experience shall be provided as specified in subsection I (1) (B).

C) A person who submits evidence of classroom or laboratory training in a vocational or trade school, a branch of the military service, or a college or university shall be given credit hours at the rate of two credit hours for each classroom hour, not to exceed a maximum of 24 months' credit.

D) Evidence shall consist of transcripts, degrees, military service records, or certificates of completion. If the course submitted by an applicant for the plumbing license examination has already been evaluated and approved by the Department, the applicant need only verify participation in the course.

If you are currently, an apprentice in a state other than Illinois the following requirements must be met prior to applying for the exam.

3) Licensed apprentice in another state or territory of the United States:

> 28. Each applicant shall have the equivalent of four years as a licensed apprentice in another state or territory of the United States.

B) Proof of practical experience shall be provided as specified in subsection I (1) (B).

C) A person who submits evidence of experience in plumbing through an apprentice-plumbing program in another state or territory of the United States, or a municipality in another state or territory, other than the State of Illinois shall be given credit on an hour-for-hour basis toward the minimum four years of apprenticeship required.

4) A person who has completed a course of study approved by the Department as equivalent to a four-year apprenticeship served by an Illinois licensed apprentice plumber:

> 29. An approved course of instruction in plumbing shall cover the subject areas and provide the number of hours of instruction and practical training specified in Section 750.550. An approved course of instruction shall total 2,800 hours of credit.

B) Evidence shall consist of transcripts, degrees, or certificates of completion to verify completion of a course that has been evaluated and approved by the Department.

A note on a few important definitions as described by the Illinois State Plumbing Code, have been listed below for clarity.

Section 750.110 Definitions

"Act" means the Illinois Plumbing License Law [225 ILCS 320].

"Agent" means an Illinois licensed plumber designated by a sponsor of an apprentice plumber as responsible for supervision of the apprentice plumber, with prior approval from the Department.

"Approved apprenticeship program," means an apprenticeship program approved by the U.S. Department of Labor's Bureau of Apprenticeship and Training and the Department under this Part, including Joint Apprenticeship Committee (JAC) Programs. (Section 2 of the Act)

"Board" means the Illinois State Board of Plumbing Examiners. (Section 2 of the Act)

"Certification" means the act of obtaining or holding a certificate of competency in plumbing inspection from the Department, pursuant to this Part.

"Certified plumbing inspector" means any licensed plumber to whom the Department has issued a certificate of competency to inspect plumbing in Illinois.

"Continuing education credit hour" means that 50 minutes of classroom time, excluding breaks, is equivalent to one credit hour.

"Course" means any class, seminar or other program of instruction in plumbing that has been approved by the Department for the purpose of complying with continuing education requirements.

"Course sponsor" means the person or legal entity who is registered pursuant to this Part and who is responsible for conducting a continuing education course approved by the Department.

"Department" means the Illinois Department of Public Health, plumbing program. (Section 2 of the Act)

"Director" means the Director of the Illinois Department of Public Health. (Section 2 of the Act)

"Governmental unit" means a city, village, incorporated town, county or sanitary or water district.

"Incompetence" means conduct in the performance of plumbing work that indicates a lack of ability to discharge the duties required to protect the health, safety and welfare of the public; failure to maintain competency in applying the standards set forth in the Illinois Plumbing Code; lack of knowledge of the fundamental principles of plumbing inspection or an inability to apply these principles; or failure to maintain competency in current plumbing inspection practices.

"Misconduct" means an act performed in the discharge of enforcement duties that jeopardizes the interests of the public, including violation of federal or State laws, local ordinances or administrative rules relating to the position, preparation of deficient or falsified reports, failure to submit information or reports required by law or contract when requested by the municipality or the Department, conduct that evidences a lack of trustworthiness, misrepresentation of qualifications such as education, experience or certification, illegal entry of premises, misuse of funds, or misrepresentation of authority.

"Retired plumber" means any licensed plumber in good standing who meets the requirements of the Act and this Part *to be licensed as a retired plumber and voluntarily surrenders his plumber's license to the Department, in exchange for a retired plumber's license.* (Section 2 of the Act)

"Revoke" means to permanently remove the plumbing license of a licensed plumber for violations of the Illinois Plumbing License Law, Illinois Plumbing Code, or this Part.

"Sponsor" means *an Illinois licensed plumber or an approved apprenticeship program that has accepted an individual as an Illinois licensed apprentice plumber for education and training in the field of plumbing and whose name and license number or apprenticeship program number shall appear on the individual's application for an apprentice plumber's license.* (Section 2 of the Act)

"Suspend" means to temporarily remove the plumbing license of a licensed plumber for violations of the Illinois Plumbing License Law, Illinois Plumbing Code, or this Part.

Section 750.310 Administration of the Plumbing License Examination

The examinations administered to applicants for a plumber's license shall be uniform and comprehensive and shall be administered in a manner prescribed in subsection (a), with the advice of the Plumbing Code Advisory Council and Board of Plumbing Examiners. The examinations shall test applicants' knowledge and qualifications in the planning and design of plumbing systems; their knowledge, qualifications and practical skills in plumbing; and their knowledge of the Illinois Plumbing Code.

The Department will provide reasonable accommodations for applicants with disabilities in accordance with the Americans with Disabilities Act. An applicant who may require an accommodation to take the examination due to a disability shall submit acceptable documentation of the disability and a proposal for accommodation to the Department at least 10 business days before the exam date.

Acceptable documentation includes a current statement or documentation from a physician licensed to practice medicine in all its branches, or a licensed chiropractic physician, in Illinois verifying the disability and providing a specific proposal for accommodation as it relates to the disability. The statement shall be on the physician's letterhead and include the address, phone number, and signature of the physician, date prepared, and the name of the applicant.

30. The examination for a plumber's license shall consist of the following:

31. Knowledge Assessment (true/false or multiple choice or fill in the blank). Questions will be based on the Illinois Plumbing Code.

2) Construction Drawings or Plans. The applicant will be required to interpret construction drawings or plans, either on paper or electronically, and complete either those construction drawings or plans or answer questions sufficient to demonstrate knowledge of plumbing fixtures, piping techniques and Code compliance.

1) Practical Application Assessment. Projects requiring the use of current plumbing techniques and materials will be completed as part of the examination. Materials will be selected from Appendix A, Table A of the Illinois Plumbing Code. The examinee will be provided a drawing and instructions for completion of each project to be assembled during the practical application assessment.

32. Each applicant will be responsible for providing his/her own tools and other required material. Each applicant will be advised in writing as to what to bring to the examination.

33. Only persons authorized by the Department are permitted in the examination area.

34. Any applicant wearing a shirt, jacket, cap, or any article of clothing bearing pictures, writing, inscriptions, or logos of any kind will not be permitted into the examination. Safety glasses shall be worn at all times when in the shop.

e) An applicant will not be permitted to leave the examination area without permission.

f) The maximum grade value of each part of the examination shall be 100 points. An applicant must make an average of 75 or above on the examination and a grade of 61 or above on each part of the examination to pass.

g) An applicant who fails to pass the examination shall be admitted to a subsequent regularly scheduled examination after filing a retake application form and fee. The application and fee shall be submitted in accordance with Section 750.300.

35. An applicant who is observed cheating during the course of an examination shall be immediately expelled from the examination in progress and that applicant's examination will be declared void.

36. The Department will send to the applicant observed cheating a notice of intent to deny the applicant's application for examination and bar the applicant from reapplying for examination for a period of not less than six months. The applicant may request a hearing, in writing, to contest the Department's notice within the time specified in the notice. If the applicant does not request a hearing in writing within the time specified in the notice, the applicant's right to a hearing shall be waived.

37. All hearings shall be conducted in accordance with the Department's Rules of Practice and Procedure in Administrative Hearings (77 Ill. Adm. Code 100).

Section 750.320 Plumbing License Examination Results

The name of each examinee and the results of the examination given to each examinee shall be recorded by written report of the State Board of Plumbing Examiners. The results of each exam shall be confidential until announced to the examinee. Upon signature by the Director, the examination results report will be considered final and approved by the Board and Department.

38. The Department shall notify each examinee in writing of the results of his or her examination.

39. The application for examination, fee receipt, examination, and other written materials deemed the Department will maintain necessary.

40. An examinee may submit a written complaint concerning the examination if he is dissatisfied with the conduct of the examination. Such complaints shall be submitted in writing and must be received by the Department within 30 days after notification of the examination results. Complaints must be factual and state the basis being used by the examinee to allege improper conduct.

41. Upon receipt of a complaint, a meeting between the examinee and the Board will be set for the same date as the next scheduled plumbing examination. The examinee will be provided an opportunity to meet with Department representatives prior to the Board meeting to attempt to resolve the complaint. A record of all such complaints and meetings shall be kept and made part of the examinee's file.

e) The examinee shall be entitled to take the examination again, at no charge, if the evidence presented before the Department and Board demonstrates that:

42. The examinee was compelled by the Department or the Board to take the test under conditions that placed him at a disadvantage in relation to all other examinees

2) That Board members or Department staff offered any special assistance to other examinees; or

3) That the examinee's test was not evaluated according to the same standard applied to the tests of all other examinees. Any grading errors by the Board or Department, discovered as a result of the review of the examinee's test, shall be corrected.

f) An apprentice plumber who has served an apprenticeship under the supervision/sponsorship of an Illinois licensed plumber and has failed the examination three times shall be called before the Department and Board of Plumbing Examiners to determine compliance with the requirements for supervision of apprentices. The employer/sponsor must accompany the apprentice at such meeting. An examinee who fails to appear shall be ineligible for admission to the next plumber's license examination and subject to license revocation. An employer/sponsor who fails to appear shall be subject to license revocation.

Section 750.400 Licensing of Plumbers

43. Initial License Examination. The Department prior to engaging in plumbing activities shall license a person. The Department shall issue a plumber's license to each qualified applicant. In order to qualify, an applicant shall:

44. Successfully pass the plumbing license examination.

45. Pay to the Department the required license fee.

46. License Renewal. All plumbers' licenses shall expire on April 30 of each year, except initial licenses issued after February 15 shall expire one year after the next April 30. A plumber's license may be renewed for a period of one year from each succeeding May 1 upon payment prior to May 1 of the required renewal fee. As a condition of renewal, a licensed plumber must provide proof of completion of four hours of continuing education in one or more courses approved by the Department.

47. License Reinstatement. *A plumber licensed pursuant to the Act whose license has been expired for a period of less than 5 years may apply to the Department for reinstatement of his or her plumber's license. The Department shall issue such license renewal provided the applicant pays to the Department all lapsed renewal fees, plus the reinstatement fee.* (Section 14 of the Act)

48. License Restoration. *A plumber licensed pursuant to the Act who has permitted his or her license to expire for more than 5 years may apply, in writing, to the Department for restoration of his or her license. The Department shall restore his or her license provided he or she pays to the Department the required restoration fee and successfully passes the examination for an Illinois plumber's license. The restoration fee includes the applicant's first examination fee. Failure by the applicant to successfully pass the plumbing license examination shall be sufficient grounds for the Department to withhold issuance of the requested restoration of the applicant's plumber's license. The applicant may retake the examination in accordance with the provisions of the Act.* (Section 14 of the Act)

Section 750.410 Licensing of Apprentice Plumbers

49. Initial License. The Department shall issue an apprentice plumber's license to each qualified applicant. In order to qualify, an applicant shall:

50. Be at least 16 years of age.

51. Verify on the Department's application form that an Illinois licensed plumber sponsors the applicant. The name and license number of the Illinois licensed plumber or program sponsor shall be included on the application form.

52. Pay to the Department the required license fee.

a) Renewal. All plumbers' licenses shall expire on April 30 of each year, except initial licenses issued after February 15 shall expire one year after the next April 30. An apprentice plumber's license may be renewed for a period of one year from each succeeding May 1 upon payment prior to May 1 of the required renewal fee. No licensed apprentice plumber may serve more than a 6 year licensed apprenticeship period.

Section 750.430 Plumbers' and Apprentice Plumbers' License Records

The Department may destroy any record relating to a plumber's license or apprentice plumber's license on which there has been no activity, such as license renewal or restoration, within the previous five years.

Section 750.900 Plumber's and Apprentice Plumber's License Violations

53. The Department may take disciplinary action against a licensed plumber or licensed apprentice plumber for violations of the Act, this Part, or the Illinois Plumbing Code. Pursuant to Section 20 of the Act, disciplinary action may include the revocation, suspension, or denial of a license issued by the Department; and under Section 5(b.10) of the Act may include an Order of Correction to a telecommunications carrier for improper advertising.

54. A violation, for the purposes of this Section, shall be considered to mean a finding of violation of a Section of the Act, this Part, or the Illinois Plumbing Code by the Director in a final order issued pursuant to the Act, and may include any one of the following acts:

55. Any licensed plumber who permits his/her name or plumbing license number to be used to imply that he/she is a member of a sole proprietorship, association, partnership or corporation, and evidence indicates that he/she is not participating in the plumbing activities of the sole proprietorship, association, partnership or corporation. Evidence used by the Department in making this determination may include payroll, records, time sheets, W-2 forms, and documents on file with the Secretary of State

56. Any licensed plumber who refuses to correct Illinois Plumbing Code violations as requested by the Department, continues to install plumbing in violation of Illinois Plumbing Code requirements, or is found guilty of negligence or incompetence in the performance of plumbing;

57. Any licensed plumber who employs individuals to install plumbing and fails or refuses to license them as Illinois apprentice plumbers

4) Any licensed plumber who fails to adequately train apprentices under his employment or supervision in a manner qualifying them to pass the plumbing license examination. In reaching a finding of adequacy, the Department shall consider factors such as number of hours worked by the apprentice, types and varieties of plumbing work performed and inspections of finished work;

5) Any licensed plumber or approved apprenticeship program that sponsors an apprentice and does not directly supervise or employ the apprentice in the plumbing trade a minimum of 1,400 hours a year

58. Any licensed plumber who does not complete continuing education as required by Section 750.500 for license renewal

59. Any licensed plumber or approved apprenticeship program that fails to submit to the Department, within 15 days after an apprentice is no longer employed and supervised by that plumber or program, a letter stating that sponsorship of the apprentice has been canceled;

60. Any licensed plumber or licensed apprentice plumber who advertises his or her services as a certified plumbing inspector without obtaining certification from the Department or who uses or attempts to use the certificate of a certified plumbing inspector;

61. Any licensed apprentice plumber who performs plumbing work without the supervision of the sponsor/agent or approved apprenticeship program sponsoring the apprentice; or

10) Any licensed apprentice plumber who does not work in the plumbing trade for a minimum of 1,400 hours a year or who does not fulfill all requirements of an approved apprenticeship program.

The Cold facts

Now that is the legal aspect of the requirements to become eligible for the exam. Here are the facts, as you already know about 90 percent of the apprentices that are out, working on a day-to-day basis, are actually not working under direct supervision. Much less supervision of any kind. Most if not all non-union service shops allow this, I know because this is exactly how I started. While on the topic of non-union, I will tell you that for the apprentices in the union have already taken the state exam. Not once or twice, but at least a half dozen or so times. The unions' train the apprentices throughout the 24 months prior to being eligible for the exam. Not that this is a bad thing, it is in fact a great thing. After all preparation is the key to successfully passing the exam.
All I am pointing out is for you non-union apprentices that have had no formal training are going to have a much more difficult time and the odds are stacked against you that you are going to pass the first time. That is another reason for this book. I am going to assume that you the reader are a non-union apprentice, well my friend I was also a non-union apprentice without formal training or exam preparation.
I learned mostly by trial and error, in some ways I was self-trained. This of course is also the school of hard knocks! Along the way, I damaged and destroyed things. However, I was never afraid of the unknown, even though I had a lot of uncertainty I continued to push on.
I had my share of callbacks; you are not alone in that. I did however have a few very good licensed plumbers that I knew I could call if I was ever in a jam. The "old school" plumbers that I knew helped me out with either a product or code that I was unsure of. In short, networking is a valuable tool of gaining information.
On the next two pages, I have included a sample of the state application form. This is the form that you as an applicant will be required to fill out. Once this is completed, you will mail it, with two pictures, and your check to the department. Remember if you do not meet the minimum requirements, the department will deposit your check. They will also inform you of your eligibility for the exam.
As you can see from 750.1100, the fees for the exam are $175 dollars. Be sure you meet the minimum requirements. I cannot stress that enough, regardless if you are or not the department will keep your check. That is another reason to pass the exam the first time. I have heard stories of people going multiple times before they are able to pass the test successfully.
As I stated earlier, this is an exam with a 70, percent fail rate. Know the material before you apply for it. Be sure that you are ready. Preparation is extremely important.

State of Illinois
Illinois Department of Public Health

OFFICIAL USE ONLY
Months Accumulated _____
Approved _____
Disapproved _____
By _____
Date _____

Application for Examination for Plumber's License
(due 30 days prior to date of examination)

Print legibly or type

Last Name	First Name	Middle Name
Home Street Address		City
State	ZIP Code	County
Work Phone		Home Phone
Date of Birth	Place of Birth (geographic location of birth)	
United States Citizen ☐ Yes ☐ No	If "no", date eligible for citizenship is _____ If "no", applicant must submit a signed and sealed N-300 or N-400 from the Department of Homeland Security.	
High School Graduate ☐ Yes ☐ No	If "no", applicant must submit documented evidence of completion of a two-year course of study in a high school or an equivalent course of study.	

Applicant's state of Illinois apprentice plumber's license number is 056- _____

NOTE: (225 ILCS 320/10), SECTION 10 (d) requires an applicant to have been employed as an Illinois licensed apprentice plumber under supervision in accordance with this act for ***at least four years preceding the date of this application.*** **It is up to the applicant to make sure he/she meets this requirement. *Application fees are NOT refundable.*** List the four or more years of employment that you have worked as a licensed apprentice plumber. Include the complete name and address of firm. Fill this out to the best of your knowledge. **Do not call IDPH for this information.**

YEAR _____	
YEAR _____	
YEAR _____	
YEAR _____	

Attach Recent
1" x 1"
Head and
Shoulders
Photograph
of Applicant

Application Continued on Opposite Side

Important Notice
This state agency is requesting disclosure of information that is necessary to accomplish the statutory purpose as outlined under the Illinois Plumbing License Law, 225 ILCS 320. Disclosure of this information is mandatory. Failure to provide any information could result in denial of the contractor license. This form has been approved by the Forms Management Center.

Printed by Authority of the State of Illinois

State of Illinois
Illinois Department of Public Health

Application for Examination for Plumber's License

TO BE COMPLETED BY OUT-OF-STATE APPLICANTS ONLY

I am a licensed plumber. ☐ Yes ☐ No	License issued by (name of licensing agency)
	Street address
	City, state and ZIP Code
A copy of my current license is attached. I completed _____ years as a licensed/registered apprentice plumber. My apprentice license/registration was issued by:	License issued by (name of licensing agency)
	Street address
	City, State & ZIP Code
Licensed by exam ☐ Yes ☐ No	Date of exam	Address/Site of exam

You must submit a letter from the licensing entity that states the exact time period that you have been licensed.

THIS SECTION TO BE COMPLETED BY IN-STATE APPLICANTS (ILLINOIS OR CHICAGO) ONLY

I have successfully completed a course of instruction in plumbing that was under the auspices of the U.S. Department of Labor's Bureau of Apprenticeship and Training. Yes No - If "yes," attach a copy of certificate of completion.

I have been supervised by the following licensed plumbers (list name and identification number of licensed plumbers/agents).

1.	
2.	
3.	

I have served approximately _____ hours in the plumbing trade under the sponsorship and supervision of the above licensed plumbers.

_____   _____   _____
(Signature of Applicant)   (Social Security # of Applicant)   (Date Signed)

FEES ARE NON-REFUNDABLE
Application Fee for Illinois Licensed Apprentice Plumber: $175
Application Fee for Plumbers Registered or Licensed in Chicago or Outside the State of Illinois: $225
Returned Check Fee: $100
DO NOT SEND CASH. Attach a check or money order, payable to the Illinois Department of Public Health.

RETURN APPLICATION WITH ALL ATTACHMENTS TO:
Illinois Department of Public Health
Plumbing Program
525 W. Jefferson St., 3rd Floor
Springfield, IL 62761
Telephone 217-524-0791 - Fax 217-524-5868
TTY (hearing impaired use only) 800-547-0466

16 EXAMINATION STRATEGIES

Test Taking Strategies

If you are enrolled in an apprentice class, be sure to attend, sounds simple enough however daily life may get in the way. What I mean by that is workload, family, appointments and so on. It is important to stay the course and attend each class that is available. I can tell you first hand when students do not show up it is as frustrating for the instructor, and unfair to those who never miss a class. This is especially true when the rest of the class has moved on to a new topic or completed a quiz. If you are the person holding everyone else up because you failed to attend a class or two. The apprentice classes I taught moved along at a fast pace. There was always time to slow down a little for those that did not understand the materials. However missing a couple of classes could mean missing very good information.

You should take your time and be sure to read and follow ALL the directions carefully. The state exam is a timed test you will have five hours to complete it. Remember you cannot bring in ANY electronic devices of any kind, I recommend you wear a watch. Be aware of how long you spend on any given portion of the exam.

Do not try to rush through it, this is not a race, finishing last is sometimes the real winner! I have included approximate times for each portion of the test. They are as follows:

Written: fifty multiple choice and true false questions related to plumbing code. Time to complete is forty-five minutes (45 minutes)

Drawing: you may have to draw the waste and vent system for a single or multiple story building. The drawing portion of the exam is the one that eight out of ten applicants will have the most trouble with. You will not know what type of drawing you will end up with so again preparation is crucial. The allotted time for this task is one hour and thirty minutes to complete. (1.5 hours)

Practical portion of the test: this is broken down into three parts consisting of a ¾-inch copper tubing project, a 4-inch single poured lead joint, and a 2-inch PVC project.
Total allotted time for the lab/or practical shop time is (45 minutes for each project) which combined is two hours and twenty-five minutes (2 hours 25 minutes). As you can see speed and accuracy is very important. Equally important is to check over your work once completed, however do not dwell over it. Check for neatness and accuracy.

Avoid cramming for a test; the human brain can only retain so much information at once. If you have already applied for the exam, without much prep work the odds are you may not pass. In other words if you are at 36 months of your apprenticeship, hopefully you have begun to study. If you have put off studying until 47 months, and have submitted the application it is already too late.

Organize a study area; to reduce interference, make sure that you have all the necessary materials (including this book!) before you begin studying. Some people are able to concentrate if they have a regular spot in which to study. I will go over the basic materials in which to focus in on later in this chapter.

Study from old exams; if you have attended apprentice class I would hope you have been given quizzes or test over each section of what you have learned or studied. Quizzes are very important; it will tell you a few things based on your scores. First, you should be able to see where your strengths and weakness are. Based on that information you will have an understanding of what areas to study more.

Exam format; as I stated above the state exam has two basic elements. One is a practical exam. This will have three sub parts including a copper project, a cast iron project, and a PVC project. I will reveal exactly what each of these are encompassing as we move further into the project section of this book. The second part of the state exam is the written portion, which includes a drawing. For this section however
I am referring to the written portion of the exam; you will have 50 questions, which will consist of true false and multiple choice.

Grading; this is crucial to understand that if you fail any one part of the exam as listed above, you will be required to take all of it over again. The exam is graded as follows; you must make an average of 75 or above and a grade of 61 or above on each part in order to successfully pass. This means that if you score a 59 on any one part, you will have to take the entire test again.

Form a study group; if you are enrolled in an apprentice class chances are that you may be going to take the exam with someone in that class. Another thing you can do is ask your sponsor. After all that is a part of his or her responsibility as a sponsor. Who better to use as a study partner than someone who has taken the exam.

Night before the exam; the one and most important thing is try and relax! In addition, avoid trying to study or cram all night. I recommend just relaxing and taking your mind off the test as much as possible. You can review or read your notes but keep it light. Eat a good meal avoid caffeine and junk foods. Above all, get a good night's sleep.

Morning of the exam; Eat a light breakfast that is high in protein, again avoid energy drinks. If you are a coffee drinker as I am limit yourself to just one cup. Too much coffee can get you jittery you will need to be relaxed and focus on the task at hand. The exam may be held at a location that will require you to stay overnight. It is a good idea to have a wakeup call rather than rely on an alarm clock.

Stay loose, relaxed, and confident, tell yourself you have trained for this and you are prepared. When it comes to attire, remember avoid any clothing that has logos company names or any other print. Dress in comfortable work clothes and boots. It is also a good idea to leave the baseball hat at the hotel or in your vehicle until after the exam.

Personal grooming; this may sound odd but remember you are actually being graded the moment you walk into the examination area. If you have, a beard or goatee keep it neat trim and professional. If you have long hair keep it tied out of the way of your face. Wear jeans or work pants that are clean and free of rips or tears. Tuck your shirt in and wear a belt. Also, leave the body jewelry at home.
Remember the proctors that hold the exam will be more likely to deduct points based on appearance as well as sloppy work. You are there as a representation of the future plumbers of this industry and state, hold yourself to a higher level as the proctors surely will. Be neat, courteous and keep to yourself. In other words, do not be the person laughing and telling jokes, your there for one purpose, PASS the exam! Take it seriously be a professional.

Strategies for Multiple Choice

When taking a multiple choice test carefully read the question fully before selecting your answer. Try to decide what the answer to the question is before you read all of the choices, but again be sure to read all the answers before selecting one. Sometimes two answers will be similar and only one will be correct. Narrow your choices down by eliminating obviously wrong answers, which are almost identical. Do not be afraid to change an answer if you feel strongly about it.

Do not be discouraged if you cannot answer a question. Leave it and move on to the next. You may find clues to the answer in subsequent questions. Beware of questions with "no", "not", and "none." These words easily change the meaning of questions. If time allows review both question and answers it is possible you miss-read questions the first time. Listed below (in no particular order) are a few guidelines – or strategies for answering a multiple-choice test.

1. Read the directions. Are you being asked to find the best response or a correct response? Answer each question in your head-first before you look at the possible answers.

2. Read the stem and all of the choices before selecting your answer. If you are not, sure of the correct answer Eliminate alternatives that are absurd, silly, or obviously incorrect.

3. Cross off answers that are clearly not correct.

4. Make sure the stem and the choice you have chosen agree grammatically.

5. Choose the alternative that is most inclusive.

6. The longest choice is usually correct. It contains elaborations necessary to make it correct. The correct choice will usually contain relative

7. Qualifiers such as usually, generally, sometimes, often, etc. These words allow for exceptions

8. The correct choice will usually not include absolute qualifiers such as always, never, at no time, etc. These words do not allow for exceptions.

9. Be alert for choices that are identical (they are usually both incorrect) or opposite (often one of them will be the correct choice).

10. Be careful when asked which is not true or all of the following apply except. These are asking you to pick an incorrect answer. If there is no penalty, guess if you do not know the answer! You have a 20-25% chance of choosing the correct response on most multiple-choice questions.

As previously stated the exam will also include true – false questions as well as the multiple-choice questions. Most of these questions will be right out of the codebook. Listed below are a few strategies for answering True – False questions.

Strategies for Answering True-False Questions:

62. Read the directions. Are you being asked to answer the question only or to answer the question and correct any false information?

63. Note how your responses are to be marked? (T/F, X/O, etc.)

64. If any portion of the question is false, the entire question is false.

65. Read carefully for names and dates that are similar and could be easily confused.

66. If you aren't sure of the correct response, keep the following hints in mind:

Longer questions are more likely to be true. Questions containing relative qualifiers (e.g., normally, frequently, most, some, etc.) are likely to be true because they allow for exception. Questions containing absolute qualifiers (e.g., never, always, none, every, etc.) are likely to be false because they do not allow for exception.

67. If there is no penalty, guess if you do not know the answer! You have a 50% chance of choosing the correct answer.

As far as guessing is concerned, you do have a fifty / fifty chance of guessing correct, however if you carefully read each question, and study using the code book, and this book as guidance you will have done your part in becoming prepared. What will I need to bring with me to the exam? A question that comes up often. On the next page I have included an example of the sheet you will receive from the state once you have filled out the exam application, and been acknowledged as eligible to take the exam.

State Plumbing Exam Materials

General Supplies include:

1. Safety glasses
2. Tape measure / ruler (folding)
3. Straight edge / or protractor
4. One blue and one green colored pencil / pencil sharpener
5. Two no. 2 pencil / quality eraser
6. Erasing shield
7. Rag / cloth

Shop practical supplies (copper / tools)
1. Solder 95/5 or lead free
2. Flux / acid brush
3. Torch in good working order
4. Sand cloth / scotch bright pad
5. Tubing cutter / pencil reamer
6. Fitting brush ¾"
7. Five feet of ¾ copper tubing hard drawn L, M, or K
8. Four ¾ 90 elbows
9. Four ¾ street 90 elbows
10. Two ¾ street 45 elbows
11. Three ¾ tees
12. One ¾ male adapter

Shop practical supplies (cast iron / tools)
1. 36" length of 4" cast iron soil pipe (with hub)
2. One five pound ingot of lead
3. Lead pot / ladle
4. Soil pipe cutter
5. Complete set of lead caulking tools including ball peen hammer
6. Oakum (brown only)
7. Soap stone or marking crayon

Shop practical supplies (PVC plastic / tools)
1. Ten feet of 2" ABS or PVC schedule 40 pipe
2. Four 2" long sweep 90 elbows
3. Four 2" street 45 elbows
4. Two 2" double sanitary wye
5. Three 2" sanitary tees
6. One 2" fitting adapter MIP x hub
7. Three 2" fitting clean out adapters with clean out plug spigot x FPT
8. One 2" cap
9. Solvent cement / cleaner
10. Tools for cutting PVC
11. Tools for chamfering PVC

No power tools or battery tools are allowed in the testing area, all work must be proved by hand tools. If you have difficulty cutting PVC pipe square it is advisable to either practice or purchase PVC ratchet cutters.

During the practical portion of the exam, you will be given a set of paperwork for each of the three projects to be completed. The most important thing I can say is read ALL of the instructions BEFORE you begin to cut, assemble, or fit any parts together.

You are being graded on the simple task of following direction. If you try to rush through the practical and forget to have a proctor grade or inspect your work prior to assembly, you will have points deducted from your overall score. Not following directions can be as devastating to your overall score as having purple primer running down the PVC project. Which by the way is not required. I however did use purple primer for my exam.

As far as the practical portion of the exam it is a recommendation to start with the PVC project first. This will allow time for the cement to become cured before the hydrostatic test. From there I would move onto the copper project. As the PVC project, remember to have your work inspected prior to fitting and final assembly. Also, remember to clean and ream all of the part for the copper project. Lastly is the lead pour or cast iron project. For this project be sure to pack the hub with oakum so that the lead will measure out to one-inch as per code. The easiest way to accomplish this is by cutting a small groove into your packing iron. Simply measure one inch from the end of the iron and cut a groove into it with a dremel tool or similar tool. This will ensure you pack enough oakum in the joint to give you a one-inch pour.

The grading is as follows for the copper project: 50 total points

Preparation: – maximum of 10 points off
Square cut: unsquared cut = 1 point off
Reaming: one or more un-reamed joint = 3 points off
Clean joint: pipe un-clean joint or pipe = 1 points off

No inspection prior to assembly: = 4 points off
Proper and approved flux and solder: un-approved materials = 1 point off

Air test: leak or leaks in project 15 points off
Appearance: possible points off = 10, this includes solder drips runs or smear. Workmanship and neatness, alignments, following directions.

Square and alignment of the project = 15 points off (not a true 8-1/2" x 12" square. Must fit into a pre-formed jig to check square.

Grading for the plastic project is the same grading structure as the copper project.

Grading of the cast iron project is as follows: 14 total points

Workmanship, failure to follow directions, yarning, caulking, and material = 3 points off
Double pour joint = 4 points off
Pour of more or less than one inch = 1 point off for every $1/16^{th}$ of an inch (over under) maximum of 4 points off
Proper tools and the use of tools = 1 point off for improper or misuse of tools.
Alignment of project = 2 points off for pipe not straight and in alignment.

Written Exam

The state of Illinois plumber's exam written portion consist of 50 questions including multiple choice and true-false. Over the next several pages, I have included many of the similar questions you can expect to find on the actual exam. Use these as a study guide for the written portion. There are a total of 200 practice questions some of which are directly from the codebook and some are math questions. It is also a good idea to have a study partner to quiz you over the course of several weeks leading up to the exam. Just reading them is not going to be enough.

I have included the answers for the sample "practice written exam" after each test. I recommend you look up the questions that you answer wrong in the codebook. This will help you to understand the question and the answer for the particular code.

Written Practice Plumbing Exam

Illinois State Plumbing Exam
Test # 1

1) What is the name of a chemical substance that does not allow oxides to form on freshly cleaned copper before soldering or brazing?

A. Solder
B. Flux
C. Cement
D. Sump

2) When a water supply system is tested with air, the pressure applied is;

A. The same normal operating pressure
B. Less than the normal operating pressure
C. Greater than the normal operating pressure
D. At a specifically calculated part of the normal operating pressure

3) An escutcheon is used:

A. On the outside of the stub out
B. On the top of the valve seat
C. In the base of a water closet tank
D. In advance of an atmospheric vacuum breaker

4) The trap seal is the maximum depth of a liquid that a trap will retain, measured between the crown weir and;

A. The dip of the trap
B. The top of the dip of the trap
C. The bottom of the dip of the trap
D. Any part of the dip of the trap

5) A vent connecting one or more individual vents with a vent stack is a;

A. Main Vent
B. Yoke Vent
C. Island Vent
D. Branch Vent

6) A pipe which makes an angle of 60 degrees with the horizontal is considered a;

A. Vertical pipe
B. Horizontal pipe
C. Crooked pipe
D. Diagonal pipe.

7) What type of joint is used in a Durham system?

   A. Sweat joint
   B. Caulked joint
   C. Screwed joint
   D. Flanged joint

8) A sewer line 50 feet long installed at a 2% grade will have a total fall of;

   A. 10 inches
   B. 6 inches
   C. 12 inches
   D. 20 inches

9) A waste pipe that discharges into and through an approved open connection is called a/an;

   A. Combination drain line
   B. Direct waste pipe
   C. Indirect waste pipe
   D. Non-pressure waste line

10) A _____ type of faucet shall be included or installed to drain water from the bottom of water heaters.

   A. Self closing faucet
   B. Weather proof lawn or frost proof faucet
   C. Boiler drain
   D. Push button faucet with 10-second delay

11) What is the fall of a line of pipe if the run is 78 feet and the grade is 1/8" per foot?

   A. 6 ¼ inches
   B. 8 ½ inches
   C. 9 inches
   D. 9 ¾ inches

12) The approved standard for gas fired storage type water heaters is listed by;

   A. IAPMO
   B. AWWA
   C. ASSE
   D. ASME

13) The _____ is located on all electric water heaters.

   A. Thermocouple
   B. High limit control
   C. Pilot
   D. Flue with baffle

14) Before mechanical devices, such as bulldozers, graders, etc., may be used for backfilling pipe trenches, the trench shall be backfilled in 6-inch layers to _____ above top crown of piping.

   A. 6 inches
   B. 12 inches
   C. 18 inches
   D. 24 inches

15) Piping and fixtures for which standards or specifications have not been adopted by Code may be installed if;

   A. Approved by the architect as an alternate
   B. Accepted by the owner as an alternate
   C. Approved by the Illinois Department of Public Health as an alternate.
   D. Manufactured by a reputable organization

16) The minimum size cleanout allowed to serve a 4 inch drain line is;

   A. 2 inches
   B. 3 inches
   C. 4 inches
   D. 3 ½ inches

17) A horizontal drain pipe has an existing load of 30 drainage fixture units. After connecting the discharge pipe of a sewage ejector pump rate at 60 g.p.m; the total load on the horizontal drain pipe will be;

   A. 90 DFU's
   B. 120 DFU's
   C. 130 DFU's
   D. 150 DFU's

18) The fitting that is required for making the yoke vent intersection with the drainage stack is a;

   A. Sanitary tee
   B. Tapped tee
   C. Double tee branch
   D. Wye branch

19) The maximum developed length for an indirect water of any sanitary waste line shall not exceed;

   A. 5 feet
   B. 10 feet
   C. 15 feet
   D. There is no maximum length

20) That portion of a fixture drain between a trap and its vent is called a;

   A. Fixture tailpiece
   B. Continuous drain
   C. Drain stack
   D. Trap arm

21) Wall mounted water closets shall be securely bolted to;

   A. The wall
   B. Double 2x6 studs
   C. A carrier fitting
   D. A closet bend

22) The overflow pipe from a fixture shall be connected on the;

   A. House or inlet side of the fixture trap
   B. Sewer or outlet side of the fixture trap
   C. Downstream side of the fixture tee
   D. Upstream side of the fixture tee

23) No shower or receptor for a single family house shall have a finished outside dimension which is less than;

   A. 24 inches
   B. 32 inches
   C. 30 inches
   D. 36 inches

24) A material not listed in the code for use in potable water piping is;

   A. Plastic
   B. Aluminum
   C. Galvanized steel
   D. Cast Iron

25) The term "Solvent Weld" applies to;

   A. Steel
   B. Plastic
   C. Copper
   D. Brass

26) Tanks flushing more than one urinal shall be;

   A. Automatic in operation
   B. Continual in operation
   C. Manual in operation
   D. Acid resistant enamel

27) The mathematical factor used by the plumbing industry to estimate the probable demand of each fixture shall be represented by;

   A. Gallons per minute
   B. Feet per second
   C. Fixture units
   D. Pounds per square inch

28) The maximum number of fixture units allowed on a 4 inch sewer with a slope of 1/8 inch per foot, as compared to a 4 inch sewer with a slope of ¼ inch per foot is;

   A. The same
   B. Less
   C. More
   D. Unlimited

29) A valve which operates by screwing a disc against a seat is called;

   A. Gate valve
   B. Ground key valve
   C. Needle valve
   D. Globe valve

30) The maximum number of water closets on a three inch horizontal branch drain line is;

   A. Two
   B. Three
   C. Four
   D. Five

31) Atmospheric type vacuum breakers shall not be subjected to;

   A. Back pressure
   B. Hot Water
   C. Minor use
   D. Intermittent use

32) In a straight run underground drainage lines over 10 inches nominal diameter; manholes shall be provided and located at intervals of not more than;

   A. 50 feet
   B. 75 feet
   C. 150 feet
   D. 100 feet

33) One of the following is not a standard fitting.

   A. Wye
   B. Double hub
   C. 1/8 bend
   D. 45 degree tee

34) The maximum DFU's allowed on a two inch horizontal waste line is;

   A. 4
   B. 6
   C. 8
   D. 5

35) Which fitting may be installed in a horizontal drain line to receive waste?

   A. Double tee
   B. Double wye
   C. Double tapped tee
   D. Sanitary tee

36) The maximum temperature of any water discharging into any part of the drainage system is;

   A. 140 degree f
   B. 160 degree f
   C. 180 degree f
   D. 212 degree f

37) The minimum constant water service pressure on the discharge side of the water meter shall be at least:

   A. 40 psi
   B. 30 psi
   C. 16 psi
   D. 20 psi

38) Of the following, the highest rate flow is;

   A. 30,000 gallons per 10 hours
   B. 1,000 gallons per hour
   C. 20 gallons per minute
   D. 1 gallon per second

ANSWER T FOR TRUE, F FOR FALSE

39) Mechanical devices shall not be installed in the plumbing vent system except for single family structures where the homeowner is doing their own plumbing work.

   A. True
   B. False

40) A combination waste and vent is permissible for floor drains only and then it must be for new construction only.

   A. True
   B. False

41) The 2004 Illinois Plumbing Code permits a water closet to have 1 ½" vent.

A. True
B. False

42) A galvanized pipe is a fitting with external threads used to close a pipe with internal threads.

A. True
B. False

43) A temperature and pressure relief valve shall discharge to the outside atmosphere if a floor drain is not accessible.

A. True
B. False

44) A saddle is permissible on a potable water supply system when installing an ice maker or humidifier.

A. True
B. False

45) Section 890 Appendix A of the Illinois Plumbing Code states that plastic water piping is acceptable for hot and cold water distribution systems.

A. True
B. False

46) Flared copper joints are permissible in an underground installation for water distribution within a building.

A. True
B. False

INDENTIFY THE TERM WHICH DEPICTS THE ILLUSTRATION MOST ACCURATLEY

47) The item below is:

A. Fixture supply
B. Fixture drain
C. Horizontal branch
D. Building zone

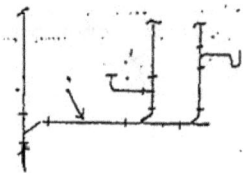

48) The item below is;

   A. An individual vent
   B. A wet vent
   C. A horizontal branch
   D. A building drain

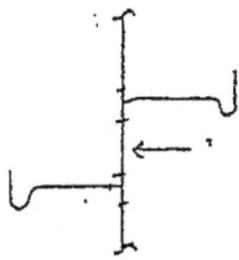

49) The vent pipe opening is above the weir of the trap in;

   A. Sketch A
   B. Sketch B
   C. Sketch C
   D Sketches A and B

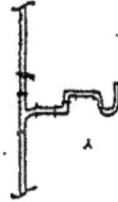 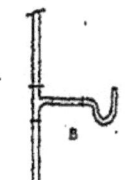 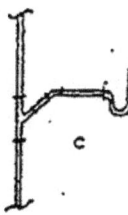

50) The connection of the food waste disposal, sink basin and dishwasher as shown is;

A. A Code violation
B. A good practice and accepted in single family homes
C. Not a Code violation
D. A Cross Connection

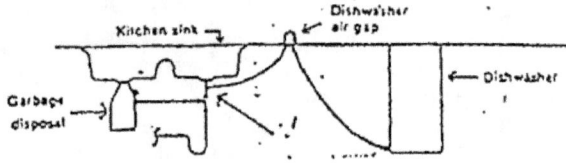

Illinois State Plumbing Exam
Test # 1 Answer Key

1) B. Flux

2) C. Greater than the normal operating pressure

3) A. On the outside of the stub out

4) B. The top of the dip of the trap

5) D. Brach vent

6) A. Vertical pipe

7) C. Screwed joint

8) C. 12 inches

9) C. Indirect waste pipe

10). C. Boiler drain

11) D. 9 ¾ inches

12) D. ASME

13) B. High limit control

14) C. 18 inches

15) C. Approved by the Illinois Department of Public Health as an alternate

16) C. 4 inches

17) D. 150 DFU's

18) D. Wye branch

19) A. 5 feet

20) D. Trap arm

21) C. A carrier fitting

22) A. House or inlet side of the fixture trap

23) B. 32 inches

24) B. Aluminum

25) B. Plastic

26) A. Automatic in operation

27) C. Fixture units

28) B. Less

29) D. Globe valve

30) A. Two

31) A. Back pressure

32) C. 150 feet

33) B. Double hub

34) B. 6

35) B. Double wye

36) C. 180 f

37) D. 20 psi

38) D. 1 gallon per second

39) False

40) False

41) False

42) False

43) False

44) False

45) True

46) False

47) C. Horizontal branch

48) B. Wet vent

49) B. Sketch B

50) A. A Code Violation

NOTES

Illinois State Plumbing Exam
Test # 3

1) Which of the following does not describe a lavatory design?

   A. Ledge Back
   B. Corner
   C. Flat Rim
   D. Receptor

2) Reverse flow in a water supply pipe caused by vacuum is called;

   A. an indirect cross connection
   B. a direct cross connection
   C. back siphonage
   D. back flow

3) The length of flush of a piston type direct flush valve is;

   A. usually externally adjustable
   B. usually internally adjustable
   C. dependent on line pressure
   D. usually not adjustable

4) The lowest piping of a drainage system which receives the discharge from soil, waste and other drainage pipes inside the walls of the building and conveys it to the building sewer is called;

   A. main soil line
   B. main drain
   C. building drain
   D. combination drain

5) Lead-free solder and flux shall be used in the soldering of;

   A. potable water lines
   B. drain, waste, and vent lines
   C. potable water lines and drain waste and vent lines
   D. all natural gas lines

6) All types of backflow/ back siphonage devices shall be field tested in accordance with the manufacture's instruction by a;

   A. certified tester
   B. pipe fitter
   C. licensed plumber
   D. professional engineer

7) A bidet shall be equipped with;

   A. hot water
   B. cold water
   C. peppermint oil
   D. hot and cold water

8) A pump on a whirlpool bath tub shall be;

   A. located above the weir of the bath tub
   B. located below the weir of the bath tub
   C. located within 7 feet of the tub
   D. located at the same height as the tub rim

9) Building drains, branches of building drains, building sewer or sanitary sewers may be located to within 10 feet of a well or suction line from the pump to the well when;

   A. cast iron pipe with mechanical joints or schedule 40 polyvinyl chloride pipe with water tight joints is used for the sewer.
   B. reinforced concrete or clay pipe is used.
   C. coil type "M" copper is used.
   D. galvanized pipe that is wrapped and coated with tar is used.

10) The minimum size of any combination waste and vent drainage line shall be;

   A. 2 inches
   B. 3 inches
   C. 4 inches
   D. It makes no difference

11) The function of a venting system is to;

   A. facilitate the flow of liquid drainage
   B. protect trap seal
   C. remove objectionable odors in the plumbing drainage system
   D. all of the above.

12) A waste pipe that discharges into a separate fixture or receptacle which when connects to the building drainage system is called a;

   A. combination waste
   B. continuous waste
   C. Indirect waster
   D. none of the above.

13) A magnesium rod is sometimes used in;

   A. storage type water heaters
   B. A flushometer valve.
   C. A T&P relief valve
   D. A closet tank

14) Compared with a grade of 1/8" per foot, a one percent grade is:

   A. considerably more
   B. slightly more
   C. slightly less
   D. exactly the same

15) All shower compartments and shower bath compartments shall be provided with;

   A. reduced pressure principle backflow device
   B. thermostatic, pressure balance, or combination controlled mixing device
   C. plastic check valve
   D. pressure gradient monitor

16) The measured length along the center line of the pipe and fitting is the;

   A. diagonal length
   B. diameter length
   C. circumference length
   D. developed length

17) The acceptable fitting for connecting a horizontal drain line with other horizontal drain lines is a;

   A. sanitary tee
   B. cross tee
   C. combination wye and 1/8 bend
   D. short sweep 90 degree elbow

18) The discharge capacity, in gallons per minute, for one unit is equal to;

   A. 7 ½ g.p.m
   B. 15 g.p.m
   C. 30 g.p.m
   D. 8 ½ g.p.m

19) The minimum size vent pipe that may serve a 4 inch trap is;

   A. 1 ¼ inches
   B. 1 ½ inches
   C. 4 inches
   D. 2 inches

20) What is the maximum distance an indirect waste for clear water be run?

   A. 15 feet
   B. 5 feet
   C. 3 feet 6 inches
   D. 24 inches

21) A water piping system with a pressure of 110 p.s.i will require a;

   A. reduced pressure backflow device
   B. extra heavy pipe and fittings
   C. pressure reducing valve
   D. pressure relief valve

22) Each plumbing fixture trap must have the minimum seal of;

   A. 6 inches
   B. 4 inches
   C. 3 inches
   D. 2 inches

23) The maximum number of water closets on any 3 inch horizontal drain shall not exceed;

   A. 2
   B. 3
   C. 4
   D. 6

THE FOLLOWING QUESTIONS ARE TRUE OR FALSE,

24) The inside diameter if any insert fitting used with plastic tubing shall not be allowed to be below the minimum allowable size for water service/distribution piping.

   A. True
   B. False

25) Where trench drains are used to carry wastes to the gas and oil interceptor, they shall have a trapped and vented opening no less than every forty lineal feet.

   A. True
   B. False

26) The minimum size of a gas and oil interceptor allowed is five cubic feet.

   A. True
   B. False

27) Water closet bowls for public use shall be of round design and the seat shall be of the closed front type.

   A. True
   B. False

28) Separate facilities for men and women are required when restrooms for the public are required.

   A. True
   B. False

29) Backflow or back siphonage devices shall not be installed where they are subject to freezing or flooding conditions.

   A. True
   B. False

30) Fire safety systems shall have a strainer on all potable water lines upstream of any backflow device.

   A. True
   B. False

31) The size of the drainage piping shall not be reduced in size in the direction of flow.

   A. True
   B. False

32) A mechanical air vent is permitted in a combination lavatory/toilet for correctional facilities.

   A. True
   B. False

33) A saddle is permissible on potable water line when installing a ice maker.

   A. True
   B. False

34) The maximum temperature if any water discharging into any part of the drainage system shall not exceed 180 degree f.

   A. True
   B. False

35) 3 7/8 subtracted from 7 ¾ is 3 7/8.

   A. True
   B. False

36) Roughing in the installation of all parts of the plumbing system that can be completed prior to the installation of the fixtures. This includes drainage, water supply, and vent piping but not the fixture supports.

   A. True
   B. False

THE FOLLOWING QUESTIONS ARE SKETCHES OF DEFINITIONS. FOR EACH QUESTION, SELECT THE WORD OR ITEM THAT FITS THE SKETCH MOST ACCURATLEY.

37)
   A. Double Wye
   B. Double Combination
   C. Double test tee
   D. Double Sanitary cross

38) A. Soil ¼ bend
   B. Sanitary Cross
   C. Combination wye and 1/8 bend
   D. Soil Wye

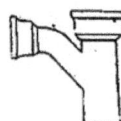

39)
   A. Soil ¼ bend
   B. Medium sweep
   C. Double ¼ bend
   D. Soil 1/8 bend

40)
   A. Running trap
   B. Soil S trap
   C. Soil P trap
   D. Drum Trap

41)
   A. Sanitary tee
   B. Test tee
   C. Partition tee
   D. Dixie fitting

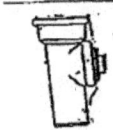

42)
   A. Soil 1/8 bend
   B. Soil 1/6 bend
   C. Soil ¼ bend
   D. Long sweep bend

43)
   A. Soil offset
   B. Long 1/8 bend
   C. Sission Joint
   D. Long ¼ bend

44)
   A. Straight tee
   B. Sanitary tee
   C. Combination wye and 1/8 bend
   D. Tapped sanitary tee

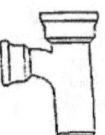

45)
   A. Double combination
   B. Sanitary cross
   C. Straight cross
   D. Triple hub tee

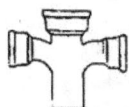

46)
   A. Bell trap
   B. Crown Vented trap
   C. Partition trap
   D. Drum trp

47)
   A. Soil P Trap
   B. Soil S trap
   C. Soil crown vented trap
   D. Soil running trap

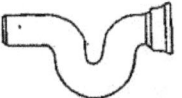

48)
   A. Soil P trap
   B. Soil S trap
   C. Soil crown vented trap
   D. Soil running trap

49)
   A. offset closet ring
   B. Fixture carrier
   C. Regular closet ring
   D. closet bend support

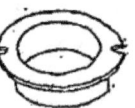

50)
   A. Cast iron No Hub drainage fittings
   B. Cast iron Durham system drainage fittings
   C. Dual-Tite compression joint drainage fittings
   D. Plastic ABS or PVC drainage fittings

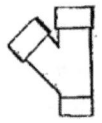 

Illinois State plumbing Exam
Answer Key Test # 3

1) D. Receptor

2) C. Back siphonage

3) A. usually externally adjustable

4) C. Building drain

5) C. potable water lines and drain, waste and vent lines

6) A. Certified tester

7) D. hot and cold water

8) A. located above the weir of the tub trap

9) A. cast iron pipe with mechanical joints or schedule 40 polyvinyl chloride pipe

10) C. 4 inches

11) B. protect trap seals

12) C. Indirect waste

13) A. Storage type water heaters

14) C. Slightly less

15) B. thermostatic, pressure balance or combination controlled mixing device

16) D. developed length

17) C. combination wye and 1/8 bend

18) A. 7 ½ g.p.m

19) D. 2 inches

20) A. 15 feet

21) C. pressure reducing valve

22) D. 2 inches

23) A. 2

24) A. True

25) A. True

26) B. False

27) B. False

28) A. True

29) A. True

30) B. False

31) A. True

32) B. False

33) B. False

34) A. True

35) A. True

36) B. False

37) B. Double Combination

38) C. Combination wye and 1/8 bend

39) D. Soil 1/8 bend

40) C. Soil P Trap

41) B Test tee

42) C. Soil ¼ bend

43) A. Soil offset

44) B. Sanitary tee

45) B. Sanitary cross

46) D. Drum trap

47) D. Soil running trap

48) B. Soil S trap

49) C. Regular closet ring

50) Plastic ABS or PVC drainage fittings

NOTES

Illinois State Plumbing Exam
Test #4

1) If you divide 16,744 by 23 and multiply the quotient by 47, the answer will be 34,216.

   A. True
   B. False

2) The pressure on the bottom of a tank of fluid varies directly with the density of the liquid in the tank.

   A. True
   B. False

3) A Glove Valve has very little resistance to flow.

   A. True
   B. False

4) Clean out plugs may be used for the installation of a new plumbing fixture.

   A. True
   B. False

5) Floor drains level with the floor shall discharge to sanitary waste drain.

   A. True
   B. False

6) A Test Tee at the base of the stack may be used as a cleanout.

   A. True
   B. False

7) A lavatory with a 1 ¼" waste outlet shall be trapped with a 1 ½" trap.

   A. True
   B. False

8) A trap shall be installed in the building drain at a point approximately five (5) feet outside the building foundation.

   A. True
   B. False

9) Flow rate is measured as volume per unit of time.

   A. True
   B. False

10) A gallon of water (U.S standard) contains 231 cubic inches and weighs 8.345 pounds (avoirdupois) at maximum density and at normal temperature.

   A. True
   B. False

11) The air gap in a potable water supply system in the unobstructed vertical distance through the free atmosphere between the lowest opening from any pipe or faucet supplying water to a tank or plumbing fixture and the flood level rim of the receptacle.

   A. True
   B. False

12) The trap seal is the vertical distance between the crown weir and the top of the dip of the trap.

   A. True
   B. False

13) The invert is the floor, bottom, or the lowest part of the internal cross section of a pipe or conduit.

   A. True
   B. False

14) Hot water is water at a temperature of not less than 120 degrees f.

   A. True
   B. False

15) Galvanized, aluminum, and stainless steel safe pans shall be at least 24 gauge.

   A. True
   B. False

16) Polyethylene, polymer concrete and fiberglass materials are not acceptable for gas/oil interceptors.
   A. True
   B. False

17) Poured concrete trench drains or poured concrete are approved for gas/oil interceptors.

   A. True
   B. False

18) Hot and cold, tempered and cold, or tempered water only shall be supplied to all plumbing fixtures that are designed for hot and cold, tempered and cold, or tempered water.

   A. True
   B. False

19) Bathtub lines/inserts shall be manufactured to an exact fit over existing bathtubs or custom fabricated according to the dimensions of an existing and shall have a slip resistance floor/bottom.

   A. True
   B. False

20) In schools that are not licensed by the Illinois Department of Children and Family Services as day care centers or homes, water closets provided for the use of children under five years of age shall be of size and height suitable for children's use, either child or juvenile type.

   A. True
   B. False

21) All self-closing faucets located in public restrooms shall be adjusted to remain open for a minimum of 19 seconds and have 1.0 g.p.m flow restrictions.

   A. True
   B. False

22) Facilities that do not have any employees working as attendants (business selling motor vehicle fuel to the public using automated machines) need not to have restrooms for public, employee, or maintenance staff.

   A. True
   B. False

23) The minimum depth for any water service shall be at least 36 inches or at the maximum frost penetration of the local area, whichever is of greater depth.

   A. True
   B. False

24) The discharge piping shall discharge indirectly into a floor drain, hub drain, services sink, sump or a trapped and vented P- trap, which is located in the same room as the water heater, for all relief valves on water heaters.

A. True
B. False

25) Reverse flow in a water supply pipe caused by vacuum is called back siphonage.

A. True
B. False

MULTIPLE CHOICE QUESTIONS

26) The lowest piping of a drainage system which receives the discharge from soil, waste, and other drainage pipes, inside the walls of the building and conveys it to building sewer is called;

A. Main Soil Line
B. Main Drain
C. Building Drain
D. Combination Drain

27) A magnesium rod is sometimes used in;

A. A closet tank
B. A flushometer valve
C. A T&P relief valve
D. A storage type water heater

28) Which of the following types of gases is lighter than air?

A. Natural gas
B. propane
C. Butane
D. None of the above

29) A peppermint test is used for testing;

A. The water supply system
B. The drainage system
C. Both water supply and drainage system
D. None of the above

30) Where is the corporation stop located in a water service line?

　　A. At the curb line
　　B. At the tap in the water main
　　C. adjacent to the water meter
　　D. Anywhere in the service line

31) The small near the top end of a dip tube in a domestic water heater serves as a;

　　A. Circulating outlet
　　B. Equalizing vent
　　C. Siphon breaker
　　D. Tempering device

32) A connection to the public sewer or private sewage disposal system is required for;

　　A. Buildings with or without plumbing fixtures
　　B. Buildings with plumbing fixtures
　　C. Buildings with emergency drain systems
　　D. Approves if accepted by owner

33) Piping, fixtures, or equipment which are installed so the it interferes with normal operation and use of windows or doors shall be;

　　A. Protected from damage
　　B. Identified from restricted use.
　　C. Relocated to avoid interference
　　D. Approved if accepted by owner

34) Cast iron pipe, installed vertically shall be supported at not less than;

　　A. Every story height
　　B. Every other story height
　　C. 5 ft. intervals
　　D. 10 ft. intervals

35) Testing water piping shall be for a period of not less than;

　　A. 5 minutes
　　B. 10 minutes
　　C. 15 minutes
　　D. 30 minutes

36) The discharge capacity, in gallons per minute, for one unit is equal to;

   A. 7 ½ GPM
   B. 15 GPM
   C. 30 GPM
   D. 50 GPM

37) The clearance required in front of a 2 inch cleanout is not less than;

   A. 6 inches
   B. 12 inches
   C. 18 inches
   D. 24 inches

38) The discharge line from S sanitary sewage ejector line shall be provided with a;

   A. Backwater valve and a gate valve
   B. Backwater valve and a globe valve
   C. Single check valve, gate valve, and a union
   D. Dual check backflow preventer and a globe valve

39) One factor for determining the size of a vent pipe is;

   A. Its accessibility
   B. Its length
   C. Type of building
   D. Price of material

40) All vent and branch lines shall be installed so as to drain back by gravity to;

   A. The main vent pipe
   B. The loop vent
   C. The center of the combination waste and vent
   D. The soil or waste pipe.

41) The maximum developed length of the indirect waster of any sanitary waste line shall not exceed;

   A. 5 feet
   B. 10 feet
   C. 15 feet
   D. 25 feet

42) That portion of a fixture drain between a trap and its vent called a;

   A. Fixture tailpiece
   B. Continuous drain
   C. Drain Stack
   D. Trap arm

43) The distance between a water heater and the union on the water lines shall not be more than;

   A. 6 inches
   B. 24 inches
   C. 3 feet
   D. 5 feet

44) No shower stall or receptor shall have a finished outside dimension which is less than;

   A. 32 inches
   B. 30 inches
   C. 34 inches
   D. 36 inches

45) Drinking fountains shall not be installed in;

   A. Patios
   B. Hallways
   C. Toilet rooms
   D. Dining rooms

46) Copper water tube type "M" shall be installed within a building;

   A. For above ground uses only
   B. Below slab only
   C. Embedded with concrete
   D. Above the ground and below slab

47) The available pressure at the water meter is 80 PSI. At the highest supply outlet, 36 feet above the water meter, the pressure would be most nearly:

   A. 44 PSI
   B. 80 PSI
   C. 116 PSI
   D. 62 PSI

48) Atmospheric type vacuum breakers shall not be subjected to;

    A. Back Pressure
    B. Hot Water
    C. Minor use
    D. intermittent use

49) The maximum number of fixture units allowed on a 4 inch sewer with a slope of 1/8 inch per foot, as compared to a 4 inch sewer with a slope ¼ inch per foot, is;

    A. The same
    B. Less
    C. More
    D. unlimited

50) The total number of water closets on a 3 inch horizontal waste branch is;

    A. 6
    B. 4
    C. 3
    D. 2

Illinois State Plumbing Exam
Answer Key Test # 4

1) A. True

2) A. True

3) B. False

4) B. False

5) A. True

6) A. True

7) B. False

8) B. False

9) A. True

10) A. True

11) A. True

12) A. True

13) A. True

14) A. True

15) A. True

16) B. False

17) B. False

18) A. True

19) A. True

20) A. True

21) B. False

22) B. False

23) A. True

24) A. True

25. A. True

26) C. Building drain

27) D. A storage type water heater

28) A. Natural gas

29) B. The drainage system

30) B. At the tap in the water main

31) C. Siphon breaker

32) B. Buildings with plumbing fixtures

33) C. Relocated to avoid interference

34) A. Every story height

35) C. 15 minutes

36) A. 7 ½ GPM

37) C. 18 inches

38) C. Single check valve, gate valve and union

39) B. its length

40) D. The soil or waste pipe

41) A. 5 feet

42) D. Trap Arm

43) D. 5 feet

44) A. 32 inches

45) C. Toilet rooms

46) A. For the above ground use only

47) D. 62 PSI

48) A. Back pressure

49) B. Less

50) D. 2

NOTES

Illinois Plumbing Exam
Test #6

1) A solar heated system shall use a double walled heat exchanger which is exposed or double walled to the atmosphere between the walls.

A. True
B. False

2) The minimum service pressure at the point of outlet discharge for all fixtures except direct flush valves shall be;

A. 8 psi
B. 4 psi
C. 15 psi
D. 45 psi

3) A pipe connecting upward from a soil or waste stack to a vent for the purpose of preventing pressure changes in the stack is a;

A. Vent stack
B. Common vent
C. Yoke vent
D. Wet vent

4) This is a condition which develops on the interior of pipe due to corrosive materials resulting in the creation of small lumps on the inner walls of the pipe.

A. Oxidation
B. Tuberculation
C. Barniculation
D. Predevation

5) The relief valve element must extend;

A. Into the to 8" of the hottest water
B. Into the side or top within 6" from the top
C. Into the dip tube
D. Into the anode rod

6) The minimum size vent through any roof is;
A. 4"
B. 3"
C. 2"
D. 5"

7) Commercial dishwashing machines and similar dishwashing equipment using hot water for sanitizing shall be provided with a sufficient supply of hot water at;

   A. 140 degrees f
   B. 180 degrees f
   C. 200 degrees f
   D. 160 degress f

8) Pressure relief valves for hot water heaters shall be set to blow of at the maximum working pressure rating of the water heater but not to exceed;

   A. 100 psi or 200 degrees f
   B. 75 psi or 200 degrees f
   C. 150 psi or 210 degress f
   D. 210 psi or 150 degress f

9) Drinking fountains or devices shall not be installed;

   A. As an integral part of or connected to any other plumbing fixtures
   B. In a penal institution
   C. In a hallway
   D. In a boiler room

10) A double fire hose connection that runs from outside the building wall through a gate valve and a check valve to the fore stand pipe is called;

   A. A true wye pattern connection
   B. An O.S.Y. valve
   C. A Siamese connection
   D. A double fire connection

11) The unobstructed vertical distance through the free atmosphere between the lowest opening from any pipe or faucet supplying potable water is called;

   A. Safe waste
   B. Air Gap
   C. Critical level
   D. Overflow rim

12) The activating handle of a flush valve in public restrooms shall be a minimum of _____ above the flood rim of the bowl.

   A. 6"
   B. 10"
   C. 20"
   D. 18"

13) Floor drains shall be trapped and have a minimum water seal of;

A. 6"
B. 2" to 4"
C. 8"
D. 1"

14) A residential dishwasher discharge;

A. May empty into a food waste disposal unit
B. May not empty into a food waste disposal unit
C. Must be connected separately into the sewer and have a separate trap
D. Must be connected to the outlet side of the trap

15) The vertical distance from the fixture outlet to the trap weir shall;

A. Not exceed 24"
B. Be a minimum of 24"
C. It doesn't matter
D. All of the above

16) The minimum size clean out plug for 4" pipe is;

A. 4"
B. 2"
C. 3 ½"
D. 3"

17) A cleanout shall be provided at, or no more than _____ feet above the base of each vertical waste or soil stack.

A. 5'
B. 4'
C. 6'
D. 1'

18) Where commercial food waste grinders are utilized, the waste shall discharge

A. Through an indirect waste
B. Into an inceptor
C. Into a separator
D. Directly into the building drainage system and not through an interceptor

19) A fixture unit drainage is the mathematical factor used by the plumbing industry as a means of estimating the probable load in the drainage system cause by discharge of various plumbing fixtures. Note: Laboratory test have shown that the rate of discharge of an ordinary lavatory with a normal 1 ¼" outlet, trap, and waste is about _____ gallons per minute.

A. 10.5
B. 7.0
C. 7.5
D. 6.75

20) No piping shall be laid parallel to footings or outside bearing walls closer than;

A. 2'
B. 18"
C. 4'
D. 12"

21) Horizontal drainage piping of three inches in diameter shall be installed with a fall of not less than _____ inch per foot.

A. 1/8
B. ¼
C. 1/16
D. ½

22) Each structure in which building drains are installed shall have one stack vent not less than _____ inches in diameter carried full size through the roof to the outside atmosphere for each building.

A. 4"
B. 3"
C. 5"
D. 2"

23) Where a roof is to be used for any purpose other than weather protection, the vent extension shall be run at least _____ above the roof.

A. 1'
B. 6'
C. 7'
D. 3'

24) Soil and waste stacks in buildings having more than ten branch intervals shall be provided with a _____ _____ at each tenth interval installed, beginning with the top floor or may be installed midway between the first and twentieth interval.

   A. Circuit vent
   B. Island vent
   C. Relief vent
   D. Wet vent

25) Where potable water supply piping must cross a sewer line below grade, a vertical separation of _____ is required.

   A. 24"
   B. 18"
   C. 12"
   D. 36"

26) The minimum size supply line for a water closet with a flush valve shall be;

   A. ½"
   B. 1"
   C. ¾"
   D. 3/8"

27) No portion of the drainage system installed underground or below a basement or cellar shall be less than _____ inches in diameter.

   A. 2
   B. 3
   C. 4
   D. 1 ½"

28. A combination waste and vent shall be permitted;

   A. Only where structural conditions precluded conventional plumbing
   B. At the installer's discretion
   C. Only if a relief vent is installed
   D. Only in storage areas and garages

29) The minimum service pressure for a direct flush valve;

   A. Shall not be less than 18 psi.
   B. Shall not be more than 8 psi.
   C. Shall not be less than 15 psi
   D. None of the above

30) All vent and branch vent piping shall have a minimum grade of _____ and connected as to drain back to the soil or waste pipe.

A. ½" per foot
B. ¼" per foot
C. 1/8" per foot
D. 1" per foot

31) What is the D.F.U for a 2" fixture drain or trap?

A. 2
B. 3
C. 4
D. 6

32) Saddles to connect ice makers for refidgerators;

A. May be used
B. Are prohibited
C. Allow only on galvanized pipe.
D. Allowed only if the homeowner install it

33) No steam pipe shall connect to any part of a drainage or plumbing system, nor shall any water _____ degrees f be discharged into any part of the drainage system.

A. Above 200
B. Below 180
C. Above 180
D. Above 140

34) When strict water main pressure exceeds 80 psi, what must be installed before the water service enters or ties into the building water distribution system?

A. An approved vacuum breaker
B. A pressure-reducing valve with a by-pass and relief
C. A butterfly flow restriction valve
D. A reduced pressure zone backflow preventer

35) The vertical distance between the crown weir and the top dip of the trap is called;

A. The outlet
B. The bottom dip
C. The top weir
D. The trap seal

36) The maximum number of drainage fixtures units on a 4" horizontal branch drain installed at a ¼" per foot slope is;

A. 150 D.F.U.'s
B. 200 D.F.U.'s
C. 75 D.F.U.'s
D. 100 D.F.U.'s

37) Plumbing systems are designed to prevent a pressure differential no greater than;

A. One-inch water column
B. 15 psi
C. 5 inch mercury column
D. 14.7 psi

38) Where is the trap located on a bath water a "P" trap us used?

A. Within 3'-6" from the outlet
B. In a wall or between floor joists.
C. Directly below the tub "waste and overflow"
D. Under the tub

39) How many drainage fixture units may be put on a 2" horizontal drain?

A. 11
B. 6
C. 8
D. 4

40) Sumps receiving the discharge of more than _____ water closets shall be provided with duplex pumping equipment.

A. 8
B. 4
C. 10
D. 6

41) In a potable water supply tank, the water supply inlet to the tank shall have a minimum air gap of at least _____ inches.

A. 2
B. 12
C. 4
D. 6

42) The maximum developed length of a dead end in a water supply system allowed is;

    A. 3 feet
    B. 4 feet
    C. 1 foot
    D. 2 feet

43) When backfilling a trench, until the crown of the pipe is covered by at least _____ of tamped earth care shall be exercised in backfilling trenches to ensure that the pipe beneath is secured.

    A. 24"
    B. 18"
    C. 12"
    D. 6"

44) For a continuous or semi-continuous flow into a drainage system, such as from a pump, ejector, or similar devices, _____ fixtures units(s) shall be considered to be equal to each gallon per minute of flow.

    A. 3
    B. 2
    C. 7.5
    D. 1

45) This type of venting only applies to floor drains and floor outlet fixtures which depend on siphonage for proper operation.

    A. Building drain
    B. Circuit vent
    C. Branch vent
    D. Island vent

46) A device used to protect the quality of water, fail safe in design, securing the potable water system by isolating a heat exchanger when the pressure between the potable water and the heat exchange medium drops below a preset level is called;

    A. Pressure relief valve
    B. Pressure gradient monitor
    C. Vacuum relief valve
    D. Sporlan pressure balancing valve

47) A vent connecting to the drain pipe through a fitting at an angle not greater than 45 degrees to the vertical is called a;

A. Vent stack
B. Waste Stack
C. Side Vent
D. Relief vent

48) Sanitary waste, gray water or mixtures containing harmful substances including but not limited to ethylene, glycol, hydrocarbons, oils, ammonia refrigerants, and hydrazine is called;

A. Sub-oil drainage
B. Toxic transfer fluids
C. Potable Water
D. Gasoline

49) What is the maximum number of water closet allowed on a three-inch horizontal waste line?

A. 4
B. 3
C. 5
D. 2

50) Adding a backflow device to a water supply system creates a/an;

A. Open system
B. Closed system
C. Semi-open system
D. Semi-closed system

Illinois State Plumbing Exam
Answer Key Test #6

1) A. True

2) A. 8 psi

3) C. Yoke vent

4) B. Tuberculation

5) B. Into the side or top within 6" from the top

6) B. 3"

7) B. 180 degrees f

8) C. 150 psi or 210 degrees f

9) A. As an integral part of or connected to any other plumbing fixture

10) C. A Siamese connection

11) B. Air gap

12) B. 10"

13) B. 2" to 4"

14) B. May not empty into a food waste disposal unit

15) A. Not exceed 24"

16) A. 4"

17) B. 4'

18) D. Directly into the building drainage system and not through an interceptor

19) C. 7.5

20) B. 18"

21) B. ¼

22) B. 3"

23) C. 7"

24) C. Relief vent

25) B. 18"

26) B. 1"

27) A. 2

28) A. only where structural conditions preclude conventional plumbing

29) C. Shall not be less than 15 psi

30) C. 1/8" per foot

31) B. 3

32) B. Are prohibited

33) C. Above 180

34) D. A reduced pressure zone backflow preventer

35) D. The trap seal

36) D. 100 D.F.U.'s

37) A. One-inch water column

38) C. Directly below the tub "waste and overflow"

39) B. 6

40) D. 6

41) D. 6

42) D. 2 feet

43) B. 18"

44) B. 2

45) B. Circuit vent

46) B. Pressure gradient monitor

47) C. Side vent

48) B. Toxic transfer fluids

49) D. 2

50) B. Closed system

NOTES

Plumbing Drawings

One of the most difficult parts of the state exam, for most is the drawing portion of the test. This can be true especially for those individuals that have not had much experience reading blueprints. For me this part of the exam was not overly challenging, simply because I had some training in reading and drawing blueprints.

The state exam can have a few different drawings for the test taker to complete. Again, it is extremely important to read the directions before starting the exam. The drawings could range from a two story, three story and a five-story building. You will be graded on neatness, accuracy, and of course proper plumbing code.

As far as proper plumbing code, you will need to know and understand drainage fixture units, and vent sizing. If you study appendix A tables E, F, G, H, I, J, K, and L. In addition, the proper use of drainage fittings is also of great importance. As with the practice test questions in the previous chapter I have also included a few basics with regards to the drawing portion of the exam. Please remember that due to space constraints I have left out the large multi-story buildings. The single drawing I have included in this book should give you a general idea of what to expect on exam day.

I have also listed a few tips and helpful diagrams to which will be of assistance in the drawing portion of the exam. Architectural drawings can be classified by in a few different ways. First you have a "plan view" which refers to the view you would have if you where to look straight down into a building as if peering in from the roof. Next, there is an "elevation" which is the view looking into a building from either the front, back or side. The state exam is based on an elevation type view.

Another example of architectural drawing as related to plumbing is called an "isometric" sometimes also known as a riser drawing. This could be a water pipe riser or waste and vent riser. An isometric drawing will consist of 30-degree angles and for most plumbers very easy to read and understand.

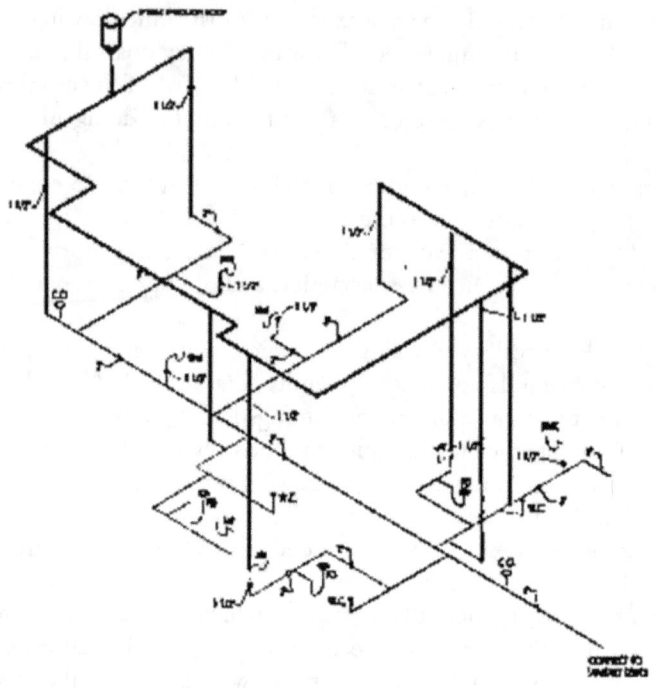

SANITARY WASTE RISER DIAGRAM

Plumbing Drawing Symbols

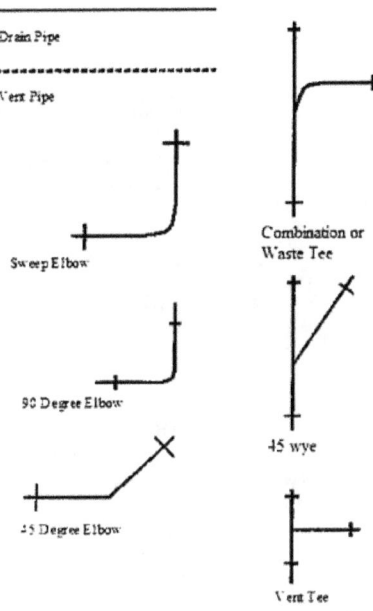

Basic fixture symbols

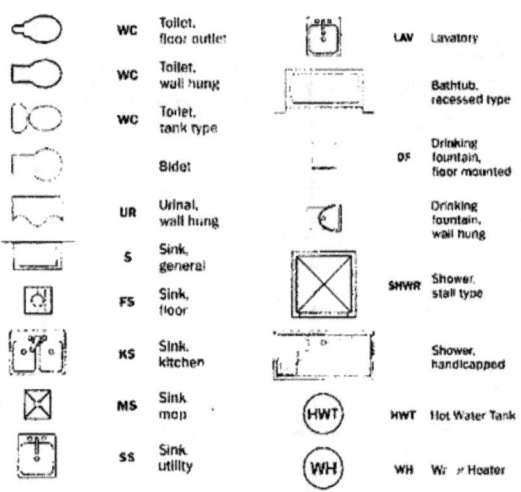

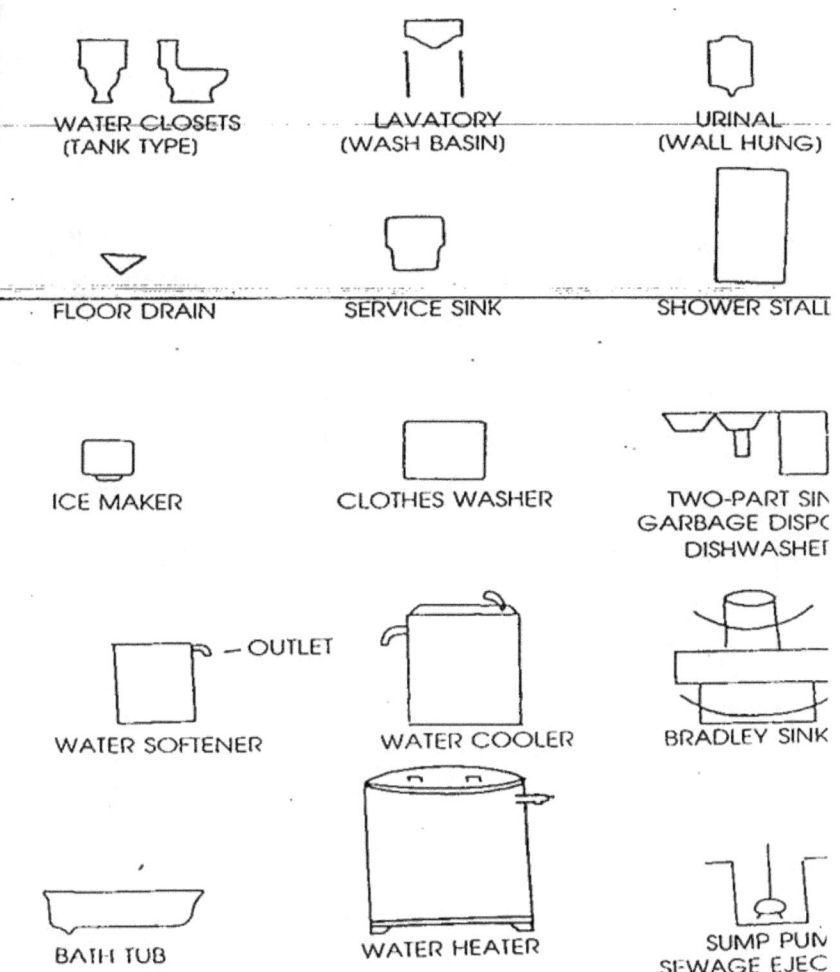

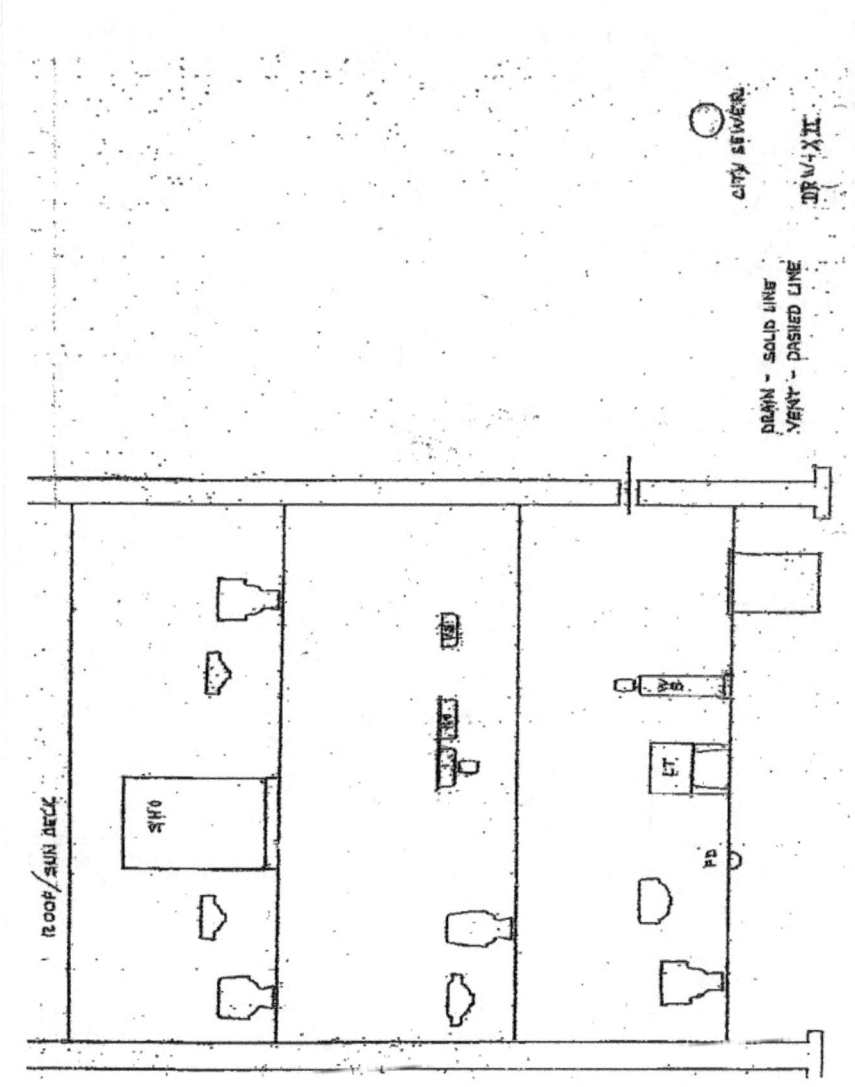

Basic elevation drawing

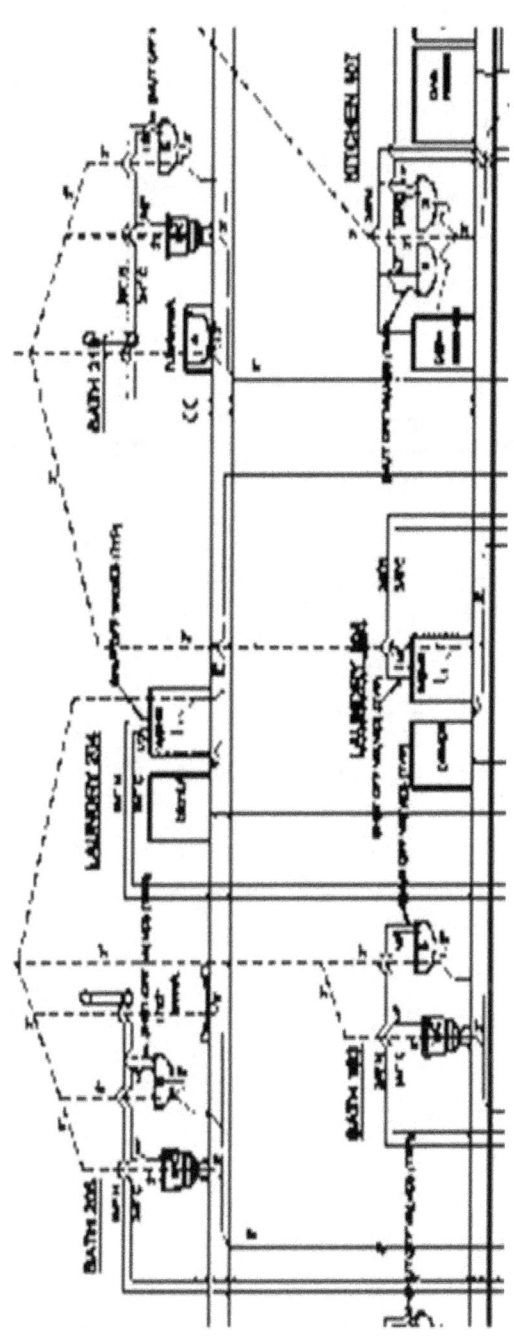

Elevation drawing with waste and vent

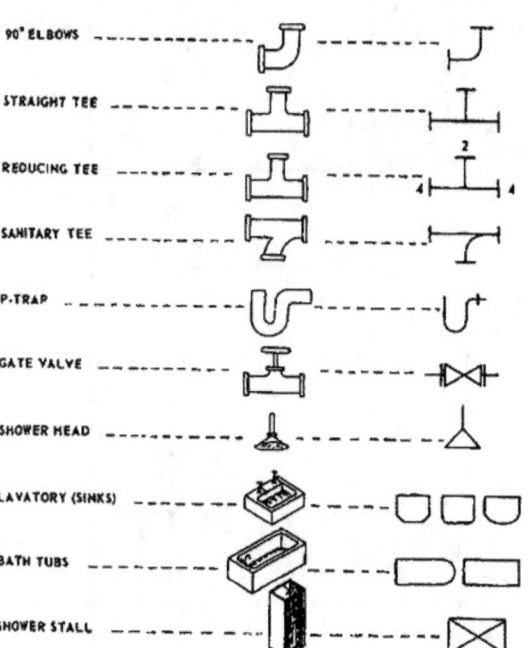

NOTES

Plumbing Math

Math may or may not be your strong suite, with that in consideration I have decided to include a section on math, as it is relative in the plumbing trade. The fact is there is a great deal of math in the plumbing industry. If we look at just a few things straight away. We can see there are calculations for just about every aspect. We have calculations for water flow, pressure, drainage fixture units, vent systems. This of course is just the few basics. Over the next few pages, I have included "cheat sheets" and charts to ease some of the pain associated with math. We as plumbers encounter daily trigonometry, algebra, multiplication, addition, and subtraction. Plumbers use math to read and calculate "take offs" to determine how much pipe will be required for a particular building or floor. Math skills are also essential to advancement within the industry.

Understanding whole numbers, fractions, and decimals and how to convert them from one form or another is necessary for the efficiency and productivity of a project. Plumbers also should be able to measure pipe and fittings quickly and accurately. This could include a 45-degree offset, or a 22-1/2 degree offset around an obstacle.

There are three pipe-measuring techniques to be familiar with. They are center to center; end to center, and face to face. These techniques are used to calculate how much pipe to cut; a skilled plumber understands which method to use based on the available information and the type of installation. Plumbers must also be able to determine fitting allowance or the distance from the end of the pipe, that goes into the fitting. Math is a very useful tool in the plumbing trades. Like your plumbing skills, your math skills will actually improve the more you use them. To keep things simple and not review basic and general math I have included a reference guide to solving most math problems in plumbing.

Basic math formulas within the plumbing trade.

Calculating one cubic foot of water:

    A. 1 CU. FT of water = 62.5 LBS
    B. 1 CU. FT of water = 7.5 GLS
    C. 1 CU. FT of water = 1728 CU inches
    D. 1 CU. IN of water = .036 LBS

Calculating one gallon of water:

    A. 1 GAL of water = 8.33 LBS
    B. 1 GAL of water = 231 CU inches

Miscellaneous formulas:

    A. 1 CU. Inch of water = .0043 GLS

B. 1 inch = .083 FT.
C. 1/8 of an inch = .01 FT.
D. 1 ton = 2000 LBS
E. 1 CU. Yard = 27 CU. FT.
F. 1 SQ. FT. = 144 SQ. inches
G. 4 inch pipe x 18 inches in length = 1 GAL.
H. 1 therm = 1000,000 BTU (British thermal units)
I. BTU = the amount of heat required to raise the temperature of 1 LB. of water
J. BTU = GALS. X 8.33 X degree rise
K. Change the decimal of a foot to inches multiple by 12
L. To find the area of a triangle divided by 2
M. To find the area of a rectangle length x width
N. To square a number multiple it by itself
O. To cube a number multiple it 3 times

P. Boiling point of water = 212 degrees

Q. Area of a circle = dia. SQ. X .7854 or PI X radius SQ.
R. Circumference of a circle PI X DIA = 3.14 X DIA.

Capacity in gallons of pipes and tanks:

A. DIA. IN. SQ. X L – IN. .0034
B. DIA. IN. SQ. X L – FT X .0408
C. DIA. FT. SQ. X L – FT. X 5.87

Capacities of rectangular tanks:

A. L – FT. X W – IN. X D – FT. = CU. IN. X 7.5 = GALS.
B. L – IN. X W – IN. X D – IN. = CU. IN. DIV by 231 = GALS

Offset Constants:

A. 22-½ DEG. = 2.61 X offset = travel
B. 45 DEG. + 1.41 X offset = travel
C. 60 DEG. = 1.15 X offset = travel
D. Travel X .707 = offset or advance for 45 fittings

Fall or rise = run X pitch per foot, pitch per foot = rise divided by run

NOTES

NOTES

Sources

Illinois 2014 State Plumbing Code 750. And 890. Appendix A including Tables Definitions, Illustrations.

Water and Health (Milwaukee 1993 – January 10, 2014)

United States Department of Labor (OSHA Section II: C-1. Domestic hot water systems)

HG.org Legal Resources Plumbing to Prevent Domestic Hot Water Scalds.

The Engineering Tool Box Sizing Water Supply Lines

Water Quality and Health Council to the Chlorine Chemistry Division of the American Chemistry Council

EPA.gov Environmental Protection Agency Flint Michigan, Detroit Free Press

History of Plumbing Uniform Plumbing Code

The Plumbing Info

Edited By my wonderful daughter Alexis Menard, Thank you for your help in making this possible. I could not have done it without you!

Useful Websites

http://www.illinoisplumbingseminars.com/
(Illinois Plumbing Seminars LLC)

http://dph.illinois.gov/topics-services/environmental-health-protection/plumbing
(Illinois Department of Public Health)

http://www.ilga.gov/
(Illinois General Assembly)

http://www.copper.org/
(Copper Development Association)

http://energy.gov/energysaver/selecting-new-water-heater
(Department of Energy)

http://www.advantageengineering.com/index.php
(Advantage Engineering)

http://www.greenplumberstraining.org/Pages/default.aspx
(International Association of Plumbing and Mechanical Officials)

https://ilphcc.memberclicks.net/
(Illinois Plumbing Heating Cooling Contractors)

http://www.americanstandard-us.com/
(American Standard)

http://www.charlottepipe.com/
(Charlotte Pipe)

http://www.theplumbinginfo.com/
(The plumbing info)

Final Word

As I complete this book, I have one main concern within our industry; and that is the lack of training by licensed plumbers. I am equally surprised at the lack of training that is offered by the contractors to its employees. Reading through the codebook in order to pass the state exam is just the beginning. Our codebook has been refined over the years. It is now a much-simplified version of what it once was. Most of the "interpretation has all but been eliminated. Yet there are still plumbers out there with the wrong attitude of getting out of their comfort zone. Just because you are "used" to, a certain way of installing something is the only way you have ever done a particular job.

Case in point, there are many myth's within the plumbing culture, that certain things such as a vent header must be at 42 inches above the floor, a vented running trap is illegal to use, or a backflow preventer is never required in a residential building. I would hear time and time again, "that's not how I was taught."

Stay educated, after all, our industry is changing quickly. Our codes are changing and developing as well. Plumbing is evolving and if you are stubborn and not willing to learn, you may end up lost.

I always tell apprentices and plumber's I have been in the plumbing industry for 30 years and I learn something new still almost every day. My philosophy on learning is the same as in business if you are not growing your dying. If you are not learning, you are stagnant as a dead end in a section of the water distribution.

Thank you for your support in deciding to purchase this book. I hope no matter if you are a licensed plumber or someone in high school on the fence about this trade, it is my hope you learned one thing: plumbing is one of the best trades in the construction industry.

I would like to say thank you to a few individuals whom without this book would have not come to fruition. First and foremost, to Ben Culos, who gave me the break in this trade so many years ago. To Danny Doyle who sponsored me under his license. In addition, to Frank Stetter who helped to guide me through the plumbing trade. Thank you guy's for all you have done to help me along the way!

www.ingramcontent.com/pod-product-compliance
Lightning Source LLC
Chambersburg PA
CBHW070312190526
45169CB00005B/1601